NITROGEN AND PHOSPHORUS
Food Production, Waste and the Environment

Keith S. Porter, Editor

New York State College of Agriculture and Life Sciences
A Statutory College of the State University
Cornell University
Ithaca, New York

A Report of an Interdisciplinary
Research Project

Robert J. Young, Director

ANN ARBOR SCIENCE
PUBLISHERS INC
P.O. BOX 1425 • ANN ARBOR, MICH. 48106

505310

Project Authors

D. R. Bouldin

H. R. Capener

G. L. Casler

J. J. Jacobs

A. H. Johnson

D. A. Lauer

R. C. Loehr

J. J. Meisinger

R. T. Oglesby

K. S. Porter

W. S. Saint

W. R. Schaffner

E. G. Srinath

R. J. Young

Other Project Participants

Research Associates and Graduate Assistants

D. R. DeLuca   C. C. Schlegel
J. L. Giles   W. T. Tseng
P. J. Godfrey   P. H. Wilde
P. H. Gore   S. Wilson
W. R. Ireson   G. M. Wong-Chong
W. H. Schaffer, Jr.

Faculty Consultants

R. W. Arnold   D. A. Haith
J. D. Francis   W. J. Jewell

Editorial Assistant

Elaina M. McCartney

Artwork

Paula Bensadoun

Acknowledgment

This report and the four year project on which it is based was supported by a grant-in-aid from The Rockefeller Foundation.

Description of Each Author

David R. Bouldin
Professor of Soil Science
Department of Agronomy

Harold R. Capener
Professor of Rural Sociology
Head, Department of Rural Sociology

George L. Casler
Professor of Agricultural Economics
Department of Agricultural Economics

James J. Jacobs[1]
Research Associate
Department of Agricultural Economics

Arthur H. Johnson[2]
Graduate Assistant
Department of Agronomy

David A. Lauer[3]
Research Associate
Department of Agronomy

Raymond C. Loehr
Director of Environmental Studies
  Program College of Ag. and Life
  Sciences
Department of Agricultural Engineering

John J. Meisinger
Graduate Assistant
Department of Agronomy

Ray T. Oglesby
Associate Professor in Aquatic Science
Department of Natural Resources

Keith S. Porter
Research Associate
Department of Agricultural Engineering

William S. Saint
Graduate Assistant
Department of Rural Sociology

William R. Schaffner
Research Associate
Department of Natural Resources

Ebar G. Srinath[4]
Research Associate
Department of Agricultural Engineering

Robert J. Young
Professor of Animal Nutrition
Chairman of the Task Force on
  Agriculture Waste
Head of Department of Poultry Science

All Departments are at Cornell University, Ithaca, New York 14853

Present Address
1. USDA, ERS
   Department of Agricultural Economics
   Cornell University   14853
2. Department of Landscape Architecture and Regional Planning
   University of Pennsylvania
   Philadelphia, Pennsylvania   19104
3. University of Georgia
   Department of Agronomy
   Coastal Plains Experiment Station
   Tifton, Georgia   31794
4. Union Carbide Corp.
   Linde Division
   P.O. Box 44
   Tonawanda, New York   14150

PREFACE

"Once the raindrop touches the surface of the earth or its appurtenances—such as rocks, trees, roofs, fences, haystacks, animals—it meets, almost immediately, abundant dust or dirt, including matter organic and inorganic, soluble and insoluble, living and lifeless. As it rolls over the dusty rock alone or as a trickling stream, it naturally dissolves some substances and sweeps on others mechanically—its departure from purity increasing as it proceeds."

Sedgwicks' *Principles of Sanitary Science and Public Health,* 1935

Man has always managed his surroundings for his own ends. In the past these have often been immediate rather than long-term. As a result, excessive exploitation of resources has sometimes been accompanied by the production of waste for which disposal was inadequate. Attitudes are now changing, and the need to conserve vital resources and the desire to protect the environment are increasingly expressed.

More stringent policies are now being applied to easily identified causes of environmental deterioration; industry, mining, and urbanization. Furthermore, a more knowledgeable appreciation of pollution and its causes has led to control policies that now include agriculture. In this respect, the aphorism of an early pioneer in the subject, "the rain to the river and the sewage to the soil" is an oversimplification of today's problems. Rain-induced flows of water whether as runoff from urban or rural areas may themselves become contaminated. Furthermore sewage or manure disposed on land can eventually enrich natural waters through leaching or runoff. This, for the eventual uses of such waters, may constitute pollution. Such less discernible non-point sources come within the province of the Federal Water Pollution Control Amendment Act of 1972. This legislation may place agriculture under significant and possibly severe constraints.

What is important for farming, if the requirements of the new legislation are to be realistically met, is that these effects must be quantified to provide a basis for proper assessment and control. In some cases, such as in food processing plants, or intensive animal units, the wastes are discharged at a point and are therefore more easily measured. But, in general, material "lost" from farming is diffuse in origin and its measurement is extremely difficult.

Coupled with the desire to preserve the environment is the objective of maximizing the efficiency of agricultural production. Given the current high costs of food and its world-wide scarcity, aims to increase production may differ from, and even be incompatible with, environmental objectives. Unfortunately the flow of water over and through the soil may remove large amounts of nutrients in soluble or particulate form. Thus the enrichment of natural waters represents a loss in potential productivity to the farmer. On the other hand, control measures may themselves be expensive and increase the cost of food production.

The studies described in this report deal with the interwoven issues of maintaining agricultural efficiency and protecting the environment. This was done by considering especially the substances nitrogen and phosphorus, both of primary economic and environmental importance. Sources of nitrogen and phosphorus found in streams and lakes have been identified. Estimates of the quantities involved have been made and their effects on lakes have been assessed with regard to management alternatives. Management of manure from treatment to direct disposal on land, and the application of fertilizer have been studied. The economic consequences of applying controls to reduce nutrient losses from farm land have also been estimated. Finally, social issues, such as public attitudes toward pollution and the efficacy of institutions responding to such attitudes, have been examined in a comprehensive sociological investigation.

To confine the investigations within reasonable limits, the group chose to primarily consider specific problems in central New York. However, many of the conclusions, such as the significance of sources of nitrogen and phosphorus, their effects, the management of animal wastes, and the economic consequences of control, have broader significance, as emphasized in Chapter 8.

To fully deal with these issues requires thorough studies of many aspects of the watershed: topography, climate, nature of the soils, geology, plant characteristics, hydrology, and the inhabitants and their activities. All these subsume the physical, chemical, biological, economic, social and engineering components of processes which are fundamental to understanding and managing the flows of nitrogen and phosphorus. Unfortunately, it is not advisable to simplify investigation by considering problems in complete isolation. Admonition of this difficulty is given by the public health inspector—perhaps apocryphal—who was quite unmoved by the rustic charms of a village that boasted neither proper sanitation nor water supply excepting a highly favored well. Upon his directive, a complete sewage system was installed, eliminating all the primitive sanitary arrangements. Unfortunately, it also eliminated the well which dried up and never yielded another drop.

Considerable efforts were made in the investigation described in this report to consider all the major ramifications of nutrient flows in agricultural watersheds. To this end, the members of the research team represented several academic disciplines—agricultural economics, agricultural engineering, agronomy, limnology, sociology, and systems analysis, all from the College of Agriculture and Life Sciences at Cornell University. The cooperative nature of this interdisciplinary venture was ambitious. As described in the Postscript to this report, occasional disagreements and different viewpoints between disciplines proved to be a sharp catalyst from which the whole group, and the investigators themselves, benefited.

This report describes a cooperative investigation and in this sense, it is a group report. In a joint effort such as this, it is difficult to properly and fairly acknowledge credit for all the work done. Throughout the investigations there was a vigorous exchange of ideas and suggestions. Also each chapter was reviewed in detail by the group as it progressed through various stages. However, for reference, major contributors to individual parts of the report are indicated at the end of each chapter.

The report has been written for both the interested layman and the scientific community. Problems discussed are also of current interest to administrators and policy makers responsible for the management of our natural resources.

To achieve readability and clarity, a serious effort has been made to clarify terms used and to ensure consistency throughout the report. To this end, a glossary is provided at the end of the report in addition to definitions within the main text. There are difficulties. A topical subject such as the environment, in which there is popular as well as technical interest, rapidly acquires a polysyllabic vocabulary liable to abuse. Such usage is illustrated by

Sir Bruce Fraser who describes a small girl pointing to her baby brother and shouting, "Mummy! Johnny's polluted his environment again" (Fraser, 1973).

The problem is not made easier when there is scientific confusion over the terms and their meaning. An important example is the term "biologically available phosphorus," which defines that fraction of total phosphorus most available to plants. Considerable care was taken in this study to distinguish this fraction from other forms of phosphorus discussed in the literature. Unfortunately, the group cannot claim to have fully resolved the ambiguity subsumed in the word phosphorus.

As will be clear from the report, it is easier to raise questions than to give their answers. Much work remains to be done. It is hoped that the following chapters will provoke and stimulate interest in the problems discussed. The group would welcome correspondence on the report or matters arising from it. Where questions concern an individual chapter, it is suggested that correspondents write to the principal author of the chapter concerned.

REFERENCES

Fraser, Sir Bruce. 1972. The Complete Plain Words (by Sir Ernest Growers, revised edition by Sir Bruce Fraser). Her Majesty's Stationery Office, London.

Prescott, S. C. and M. P. Howood. 1935. Sedgwick's Principles of Sanitary Science and Public Health. Macmillan Co. New York.

CONTENTS

1

Nitrogen and Phosphorus
in the Environment

Chapter 1

NITROGEN AND PHOSPHORUS IN THE ENVIRONMENT

Now from all parts of swelling kennels flow
And bear their trophies with them as they go,
Filth of all hues and odors seem to tell
What streets they sailed from by their sights and smell;
Sweepings from butchers' stalls, dung, guts and blood,
Drowned puppies, stinking sprats, all drenched in mud,
Dead cats and turnip tops come tumbling down the flood.

—Jonathan Swift (1667–1745), describing effects
of a heavy shower on the streets of London.

INTRODUCTION

The investigations described in this book have two main themes. The first is to assess the effects of man's activities, especially in rural areas, on the circulation of nitrogen and phosphorus. Included in this assessment are economic and sociological factors related to management and policy. The second theme is to consider management of nitrogen and phosphorus which maintains desired levels of agricultural production without degrading the environment.

These themes reflect two concerns of primary importance for modern man: to conserve resources and to maintain or improve the quality of the environment. In the United States, preservation and improvement of the environment has become a major national goal, as embodied in the 1972 Amendments to the Federal Water Pollution Control Act, PL 92–500.

This chapter introduces the investigations in this book by briefly describing a) why the problems investigated are important, b) what the causes and effects of the problems appear to be, and finally c) how the investigators approached the study of these problems.

The nitrogen and phosphorus contained in fertilizers and animal

3

and human wastes are actual or potential resources. However, as described in later chapters, the use of fertilizer under certain conditions can adversely affect the quality of ground and surface waters. Similarly, the disposal of organic wastes may affect the environment by degrading the quality of water resources, or that of the air (by the release of ammonia and other odor-producing gases). Water and air are both resources, and in this sense, preservation of the environment and conservation of resources are one.

The use of chemical fertilizers is a relatively recent development, but as Klein (1962) states, the "disposal of human wastes and other organic refuse without creating a nuisance has been a problem since time immemorial." Some ancient societies were strict in their sanitary codes, as illustrated by the Mosaic injunction to "turn back and cover that which cometh from thee" in the soil (Deut. 23:13). Given sufficient available land, disposal in this manner is potentially useful. If plant residues, and the wastes from animals and humans consuming the plants, are all returned to the original soil, then a closed cycle is preserved and nutrient additions are unnecessary since none are lost.

The use of excrement to increase the growth of plants has a long history. Sir John Russell (1971) cites a 12th century Moor living in Seville who gave explicit directions for the making of compost from organic waste. The Moor recommended the addition of blood, especially human blood, for best results. Also, the Englishman Plot (1677) highly recommended "old rags" discarded by men and women which were "well sated with urinous salts contracted from the sweat and continued perspirations attending their bodies." Apparently farmers preferred garments discarded by the poor because of their greater perspirations. Farm workers were less enthusiastic because of the fear of smallpox. The Chinese, in particular, for centuries maintained high levels of soil fertility by using their wastes. William Hinton (1970) reported that in pre-revolutionary China, a farmer's most valued possession was the contents of his privy. Farm workers were required to use their master's privy while at work, and peasants were known to locate their privies near frequently travelled paths hopeful that passersby would leave "deposits."

In Western societies, this tradition was broken as man increasingly congregated in large cities. Contents of chamber pots were often simply hurled through upstairs windows to the obvious peril of persons in the street below. Following the industrial revolution, wastes were increasingly discharged into rivers, a process accelerated by the installation of sewers in many cities in the 19th century. Outbreaks of waterborne diseases such as cholera and typhoid in major cities were frequent, and development of methods of water and wastewater treatment were primarily intended to control such diseases.

As wastewater treatment plants were more widely installed in the first half of the 20th century, and epidemics became less frequent, more attention was given to the problem of organic pollution in rivers. Since World War II, the adoption of biological treatment in sewage works has become general in the United States. This reduces easily degradable organic substances to inorganic compounds before discharge into surface waters. These compounds may serve as nutrients, and as Sawyer (1965) states, as a "result, aquatic areas have often become fertilized beyond desirable levels."

This trend in the management of wastewater has promoted several questions over the past two decades. First, there is the shift in concern from organic to inorganic substances in surface waters. Some justification for this is given by a U.S. Environmental Protection Agency (1974) study of trends in the quality of the nation's major waterways. The study showed that whereas the waterways are improving with respect to coliform bacteria, oxygen demand and dissolved oxygen, there were general increases in the recorded concentrations of nitrate-nitrogen and total phosphorus.

Second, the reduction in gross pollution from industrial and domestic wastes has transferred attention from substances in point source discharges to those originating from non-point sources such as farm land. In 1967, a Task Group sponsored by the American Water Works Association reported that "the quantity of nitrogen and phosphorus in rural runoff appears to be greater than that contributed in domestic wastes. Agricultural runoff is the greatest single contributor of nitrogen and phosphorus to water supplies" (Task Group Report, 1967). Estimates of contributions of nutrients from various sources, provided by the Task Group, indicate that up to about 60% of the nitrogen and 40% of the phosphorus in water supplies might originate from agricultural land. Other published evidence has suggested that agriculture is a major source of contaminants in rivers. For example, the U.S. Department of Agriculture (1955) estimated that approximately 1 billion tons of sediment were transported within the United States to sea every year, a large part of which was attributed to agriculture.

A third question concerns the effect of the nutrients on receiving waters. In a survey conducted by the Task Group cited above, it was found that more than half the surface waters in the United States used as water supplies were apparently affected by problems caused by excess nutrients (Task Group Report, 1966). The enrichment of these waters produced unwanted algae and state health departments indicated that 43% of impounded water supplies were sufficiently affected to require the application of chemicals such as copper sulfate. In California, more than 95% of surface waters impounded for water supply were treated with algicidal chemicals as a matter of routine.

Fourth, another major issue is the very rapid increase in the use of phosphorus in detergents since World War II. It was estimated in 1967 that 13% of all manufactured phosphatic compounds were used as "builders" to improve the effectiveness of detergents (Task Force, 1967). As a result, the amount of phosphorus discharged in wastes to surface waters has approximately doubled. Since 1972, controls have been imposed on the use of phosphatic detergents in Canada and in some parts of the United States. Although it is premature to reliably assess the effects of such prohibitions, it is believed that the effect of reduction of this one source of phosphorus is potentially greater than that of any other single source of phosphorus in surface waters (Vallentyne, 1974).

Finally, a fifth question concerns the loss of resources that the discharge of nutrients to surface water represents and means for their conservation. For example, as Sawyer (1965) has argued, the "trend of man to live in urban rather than rural areas has caused a tremendous dislocation of phosphorus distribution within the environment. Because of the custom of using water as a carrier for human wastes much of the phosphorus, as well as other fertilizing elements removed from the soil, is transferred to rivers, streams and lakes."

As the magnitude of such losses has become more evident, so has the possibility of recovering part of the nutrients been considered. Apart from recovering nutrients during treatment, another technique to recover or at least reduce their loss is to apply the wastewater to land, rather than discharge it into rivers (Sullivan et al., 1973). In a sense, it is foreseeable that management of wastewater intended to recycle nutrients would constitute a return to ancient methods in principle if not in practice.

The dual problems of conservation and water quality were central to the investigations described in this book. As will be outlined later, effects of inorganic forms of nitrogen and phosphorus on surface waters were quantified, sources of the nutrients were identified, the relative contribution from farm land under different uses were estimated, and losses of nutrients and some methods of control were determined and assessed.

SOCIAL AND ECONOMIC BACKGROUND

Associated with the technological advances briefly sketched in the previous section were social and economic trends which have culminated in what has been termed the "New Environmentalism" (Council on Environmental Quality, 1973). As stated by the Council on Environmental Quality (1970), the quality of water inspires strong public concern. "First, the growth of industries and cities has multiplied pollution in most waterways; second, demand for outdoor recreation has grown in a society increasingly affluent and leisure oriented; and third, a thread running through all the others— is man's inexplicable affinity to water." This affinity has recently found effective expression in many social groups and organizations which have promoted improvement of the nation's waterways, lakes and environment. In a democracy such promotion can be directed by the enactment of laws which define and regulate the practices considered to be undesirable. This is certainly true of efforts to preserve and restore the nation's rivers and lakes.

The path of progress is never smooth, however, and in the United States there are many groups with conflicting interests which rarely exhibit unanimity over the nation's welfare as it relates to environmental issues. To promote legislation, therefore, education is required—which when applied to legislators is less euphemistically called lobbying. Once enacted, regulations must be accepted by a majority of the members of society and compliance must be enforceable on the minority to whom they are less acceptable.

Campaigns which have promoted the acceptance of ideas and, in turn, policies governing the environment in the United States have engaged the efforts of many distinguished men in its history. Included among their number are Thomas Jefferson, Ralph Waldo Emerson, John James Audubon and Henry David Thoreau (Council on Environmental Quality, 1972). The influence of these men was initially more philosophical than it was political, and over specific issues philosophical arguments were unavailing when pitted against those based on costs. An example was the long drawn out controversy in the first decade of this century over the Hetchy-Hetchy Reservoir in Yosemite National Park which was unsuccessfully opposed by John Muir, the founder of the Sierra Club. Since then, the Sierra Club and other conservationist groups have won increasing public support with correspondingly greater successes. A large measure of this success is attributable to the increasing employment of scientific and economic arguments to support environmental causes.

An important economic concept now widely applied to environmental economics is that of external economies and diseconomies. External economies are defined to be the favorable consequences for individuals or groups resulting from the activities of others. Conversely, an external diseconomy is the harm, or disbenefit, inflicted consciously or otherwise by the action of one person or group upon another. The evacuation of a chamber pot through a bedroom window was an internal economy for the would-be sleeper, but to the passerby who got drenched, it was an external diseconomy. On the other hand, discontinuation of sewage discharge to a lake may be an external economy to swimmers and to those using the lake for water supply.

Such concepts and environmental problems were widely discussed during the 1960's in a veritable deluge of books, articles, campaigns and advertising (the Sierra Club even lost its tax deductible status after inserting a full page advertisement in the *New York Times*). A great deal was achieved both locally and regionally. In 1972, it was reported that there were at least 200 major court cases, within a two-year period, initiated by citizens concerned with environmental problems (Cahn, 1972). Nationally, the swell of public support for solutions to environmental problems was reflected in Federal expenditures on general measures to control pollution, which increased annually from 1969 to 1973 at an average rate of 26.6%.

The efforts of environmental groups were crowned by the enactment of the Federal Water Pollution Control Act Amendment of 1972. This act has been fraught with dissension and controversy in its lengthy drafting, in its eventual passage despite a Presidential veto, and now in its application. It probably represents the most stringently sweeping attempt to eliminate pollution ever placed on the statute books. Through the Act, controls will be imposed on all "point source" discharges, which must provide the "best practicable" or "economically achievable" methods of treatment by 1983. An escape clause, however, is available for discharges where it can be shown that "there is no reasonable relationship between the economic and social costs and the benefits to be obtained" (Sec. 302). Given the uncertainties in estimating benefits such as external economies, and lacking a generally accepted method of economic analysis for their determination, this escape clause may be widely exercised.

A requirement of the Act which could have considerable impact for agriculture is that states must develop area-wide plans to identify non-point sources of pollution including runoff from land used for animal and crop production, and to establish ways "to control to the extent feasible such sources" (Sec. 208).

The Act is both long and very detailed. If fully implemented, there will be drastic changes in the American scene.

Unfortunately, the Act in some of its provisions does not convincingly show an appreciation of why present practices exist or how, if at all, they can be changed. The vagueness about the means whereby some objectives can be attained incurs doubt about the Act's total viability. This is especially true of those provisions which apply to agricultural runoff and diffuse sources of pollution such as contaminated ground water.

BACKGROUND RELATIVE TO AGRICULTURE AND THE ENVIRONMENT

There are several major ways that agricultural production can affect the environment. In the 1960's, problems associated with agricultural pesticides became known to the point of notoriety following the publication of "Silent Spring" (Carson, 1964). Less publicized than pesticides were the waste discharges from numerous canning, food processing and packing plants. The major concerns of this book, however, are other means whereby farming, i.e., cropping practices and animal wastes, may affect the environment, especially the hydrosphere.

As already stated in a previous section, agricultural land has been identified over recent years, although perhaps inconclusively, as a major source of nitrogen and phosphorus occurring in surface waters. Concerning animals, it was asserted by a report of the President's Science Advisory Committee that "the excreta of farm animals are a major source of water pollution, entering streams, rivers or lakes either in surface runoff or through underground seepage, and posing hazards to human and animal health from pathogens common to animals and man. In some waters, pollution from farm animal manure has caused fish kills; elsewhere it has resulted in extensive damage to oyster and other commercial shell fisheries" (Environmental Pollution Panel, 1965). Many studies tended to confirm these assertions. Stewart (1970) cited studies showing large accumulations of nitrate-nitrogen under beef feeding lots. Measurements of nitrate and ammoniacal nitrogen in water samples adjacent to lots had concentrations ranging from 1.1 to 31.0 mg/l for nitrate and 5.1 to 38.0 mg/l for ammonia.

Figure 1.1. Runoff from dairy cattle holding area.

Farmyard manure has always, of course, been a potential source of pollution. The major respect in which present conditions differ from those in the past is that over the recent years there has been a very rapid trend to produce livestock in larger and fewer farm units located in particular geographical areas. This trend has been mainly impelled by the need to use labor more efficiently, and to obtain economies of scale which permit greater profitability. For example, Loehr (1974) cites that whereas in 1968 there were 24,000 dairy farms in New York State with approximately 40 cows per farm, by 1985 it is estimated there will be 10,000 farms with the sizes of the herds approximately doubled. The increase in size of beef feedlots and their impact is especially evident. Although feedlots with capacities for 1,000 cattle or more represent less than 2% of all those in the United States, they now produce more than 50% of all beef. Similar trends are evident for other livestock such as in poultry and hog production. Trends in poultry production are vividly shown in Figure 1.2. The percentage of farms with chickens has very dramatically fallen, especially since 1950, while the size of flocks in poultry farms has soared in the same period.

The economies of scale combined with the relatively low costs of transporting material around the country have conferred eco-

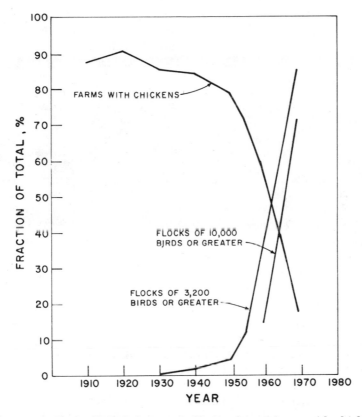

Figure 1.2. Chronological changes in the number of farms with chickens and
the sizes of flocks in the U.S. (Jewell, W. J., *et al.,* 1974).

nomic advantages on the large units. As a result, in recent years the
emerging pattern of agricultural production has been as follows:
(a) chemical fertilizers are imported to those areas producing grain
for feed, and (b) the grain is then transported to the livestock units
from whence the animal produce is shipped to the centers of
population.

In general it would appear, therefore, that the cycle of nutrients
from soil to plants to animals and to humans and then back to the
soil has in many cases been broken. The pertinent question now,
especially with regard to PL92–500, is whether the economic factors that have encouraged the intensification of animal production
might be modified by costly requirements for the treatment and
disposal of the extremely large volumes of manure produced in
big feedlots. The cost of waste disposal coupled with increases in

Figure 1.3. Dairy cattle by a small stream—a diffuse source of waste.

the cost of energy and the attendant increases in the cost of transportation will be a possible disadvantage for concentrated animal
production.

A corollary aspect of management is that farmers have used inorganic fertilizers rather than manure to maintain soil fertility,
because fertilizers have been relatively cheap compared to the cost
of applying the manure to land. Here also relative costs may change
as PL92–500 is implemented. Under the provisions of the Act requiring the control of runoff from feedlots, the farmer may have no
alternative but to apply the manure to land regardless of relative
economies in the future.

With respect to the use of inorganic fertilizers, the situation is
even more complicated. As with the trend in the size of feedlots,
the increases in the uses of fertilizer have also been spectacular. In
the 20-year period between 1950 and 1971, the use of nutrients in
fertilizers has approximately tripled and is currently about 9×10^6
metric tons per year (Hargett, 1974). This is equivalent to about 60
kg of nutrients per harvested hectare, or 40 kg per person per year.

In the reports quoted earlier both Sawyer (1965) and the Task
Group Report (1966) drew attention to the rapid increase in the
amounts of fertilizer applied to agricultural land. Although no direct
evidence is presented, the inference is that this increase in part
explains the levels of nutrients observed in water supplies. In their
later report (1967), the Task Group asserted that agricultural runoff
was the greatest single source of nitrogen and phosphorus in surface waters used for water supply. More recent reviews (Com-

mittee on Nitrate Accumulation, 1972, and Loehr, 1974) tend to confirm that agriculture is a major source of nitrate-nitrogen in ground and surface waters. It is not known with certainty, however, what amount of nutrients would be lost from farm land if fertilizers were not applied. Clearly, only the difference between losses from fertilized fields and those that would naturally occur should be attributed to the fertilizers.

In discussions on the use of chemical fertilizers in agriculture, it is sometimes overlooked that farmers have an economic incentive to make large applications to their crops because fertilizers are relatively inexpensive. Farmers will apply fertilizer as long as the value of the increase in harvest exceeds the cost of additional fertilizer. The marginal return decreases with each additional increment of nutrient, but if the marginal return from the crop exceeds the marginal cost of the fertilizer, then on purely economic grounds so far as the farmer is concerned, the additional amounts should be applied. A statistical analysis of the use of nitrogenous fertilizer on corn and the effect on yield has been reported by Jacobs et al. (1973). Results of this analysis indicated that increased losses of inorganic nitrogen to the atmosphere and to ground water can be expected following higher rates of application of nitrogenous fertilizer.

Finally, in the current context of concern about the environment, limited resources could mitigate against very heavy uses of fertilizer in the future. Justus von Liebig, the eminent German scientist of the nineteenth century, argued that since 1 metric ton of farmyard manure contains only about 15 kg of major nutrients, it was therefore more effective and cheaper to apply these in the form of inexpensive salts. There has been a great demand for such "salts" ever since. At the time of Liebig a source of phosphate was ground bone. As Sir John Russell (1971) relates, when the Englishman John Bennett Lawes demonstrated that sulfuric acid improved the use of bones as fertilizer, the demand for bones was such that Liebig angrily declared that British dealers were "ransacking the battlefields of Europe." It is arguable that man today is ransacking his present supplies.

In summary, the above discussion has outlined how the natural flow of nutrients from soil to man and back again has been broken. Large increases in population and urbanization have made the conservation of resources and environmental quality more critical. As the immediate effects of urbanization and industrialization on water quality are being dealt with, the effect of agriculture has become more a matter for concern. Thus the problem of agricultural pollution has only recently emerged in its entirety. Existing relevant knowledge is far from sufficient for understanding the problem and for managing it. The objectives of the investigations described in this book were specifically addressed to these issues: to further scientific understanding of the flow of nitrogen and phosphorus, especially in the production of food and consequent wastes, and to establish appropriate guidelines for their management.

NITROGEN AND PHOSPHORUS
WITH RESPECT TO THE WATERSHED

A fundamental factor in generating and maintaining the flow of nutrients in the environment is the flow of water, or hydrological cycle. This cycle may conveniently be considered in terms of a watershed; a drainage basin for a stream or lake.

A watershed is a basic unit of analysis broadly determined by the hydrological cycle (Figure 1.4). The flows of water depicted by

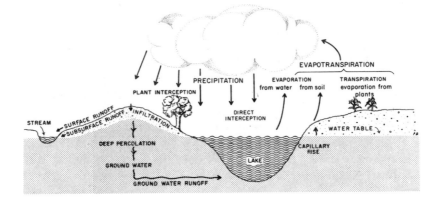

Figure 1.4. The hydrological cycle and the watershed.

arrows to the stream or lake, respectively, define the watersheds of each. The same arrows could also represent the flow of nitrogen and phosphorus to a receiving water such as the stream or lake. Surface runoff, and to a lesser extent subsurface flows, can convey both organic or inorganic nitrogen and phosphorus in either soluble or particulate form. Deep percolation and ground water runoff to a receiving water generally carry only soluble substances since the soil acts as a filter for the percolating water. Such ground water is far more likely to contain higher concentrations of nitrogen as ni-

trate than phosphorus. Nitrate is highly soluble, whereas phosphorus is readily adsorbed by soil particles and hence relatively
immobile. However, since surface runoff can move and carry soil
particles, it can thereby transport phosphorus in much larger concentrations than deep ground water. The amounts of nutrients so
moved depend on many factors including the volume of flow of
water, the characteristics of the soil and subsoil, and no general
figure can be given. For example, in arid climates there is insufficient water to induce the flow of nutrients and these accumulate
with other salts in the soil. However, Biggar and Corey (1970) state
that the concentration of dissolved inorganic phosphorus in soil
solution is generally less than 0.2 mg/l and a range of 0.01 to 0.1
mg/l is usual.

MAJOR EFFECTS OF NITROGEN AND
PHOSPHORUS ON RECEIVING WATERS

Some forms of nitrogen and phosphorus, such as nitrate-nitrogen
and soluble phosphorus, are readily available to plants. If these
forms are released into a receiving water, then in effect this water is
fertilized and abundant growths of aquatic plants can result. This
process of enrichment is called eutrophication. The flows of water
and nutrients in a watershed clearly occur naturally, and eutrophication can likewise occur in the natural course of hydrological
events. However, man's activities can substantially increase the
amount of nitrogen and phosphorus released into surface waters,
the resulting enrichment being called cultural eutrophication.

Enrichment is usually a far more significant phenomenon in lakes
than in rivers because lakes can retain the nutrients for a much
longer period, and they are therefore available for aquatic plants,
such as algae. If algae are in sufficient numbers, the clarity or transparency of the water in the lake may be markedly reduced. More
serious consequences following excessive algal growth can be the
death of fish in the lake and the increased cost of treating water
when the lake is used as a water supply (Gamet and Rademacher,
1960). Such consequences are often conspicuous and unpopular,
and eutrophication has become a major topic of interest to scientists and the general public alike.

The flow of phosphorus into and within a lake is represented in
Figure 1.5. As can be seen from the figure, the total phosphorus discharged by a stream is separated into dissolved phosphorus which
remains in the water column as shown, and the phosphorus in par-

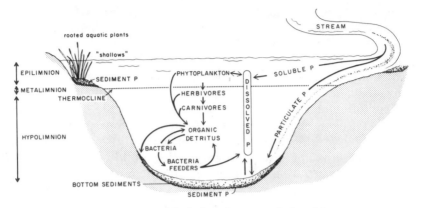

Figure 1.5. Simplified phosphorus cycle in a lake.

ticulate matter which settles to the bottom under the relatively quiescent conditions of the lake. That part of the soluble phosphorus which is accessible to the biota of the lake may be utilized by various organisms and their predators. Most of this utilized phosphorus eventually settles to the bottom of the lake as the organisms die. Some phosphorus is recycled from the sediment but on balance it is believed there is a net loss of phosphorus from the water column to the lake bottom. This suggests that if the supply of phosphorus to the lake can be reduced, then the concentration of phosphorus in the lake waters will correspondingly fall despite the potentially abundant supply in the sediment.

There has been considerable debate over which nutrient primarily induces algal growth. With respect to temperate lakes in the United States, it appears that phosphorus is probably the crucial substance. From work begun in the 1930's it has become increasingly apparent that certain algae are capable of directly obtaining nitrogen through chemical fixation, in which case nitrogen is not limiting (Hutchinson, 1957).

The presence of nitrate-nitrogen in water may be a matter for concern when consumed by animals or humans. In the stomach, nitrate can be converted to nitrite-nitrogen on rare occasions; this in turn is absorbed into the blood stream where it reacts with hemoglobin to form methemoglobin which reduces the capacity of the blood to carry oxygen (methemoglobinemia).

In humans, the incidence of methemoglobinemia almost entirely occurs in infants less than three months of age. The standard of 10 mg N/l as NO_3 is based on the incidence of methemoglobinemia in this age group. Generally, the correlation between nitrite concentrations and the occurrence of methemoglobinemia is statistically weak, especially in subclinical cases in which there are no overt symptoms. Animals are also subject to methemoglobinemia although again the evidence is not incontrovertible.

SOURCES OF NITROGEN
AND PHOSPHORUS DUE TO MAN

On the basis of the hydrological cycle, there are two main ways in which human activity can increase the flow of nutrients from land to water. First, the transport of nutrients in surface runoff and in ground water can be increased by the following: (a) the disturbance of the soil mantle or its vegetative cover, such as tillage of soil and construction of highways and building, and (b) addition to the supply of nitrogen and phosphorus on land, such as application of fertilizer. Second, the transfer of nutrients is greatly increased by "artificial drainage" such as sewage, drains in cities, and ditches in rural areas. In this case, release of nutrients to surface waters can be directed.

The above distinctions are in some cases arbitrary. For example, it is impossible to so classify septic tank systems, a consideration which is highly relevant to this study. Septic tanks and associated disposal fields can affect the quality of ground and surface waters. The Task Group cited previously (1966) reported a survey in Minnesota which indicated that leaching from these systems significantly increased the concentration of nitrate-nitrogen in wells from which homes obtained their water supply. It is also probable that many tanks "leak" into drains, ditches or directly into lakes and rivers. In such cases both phosphorus and nitrogen will be transferred from the tanks to surface waters.

Of equally uncertain significance is the transfer of nutrients to the atmosphere. Two important examples in which man may be a principal agent are wind erosion of soil and the release of nitrogenous gases such as ammonia to the air. Both the soil particles and ammonia are probably returned to the earth via precipitation, but the total impact of this deposition and how it is affected by human activity is not known (Chapter 4).

OUTLINE OF THE STUDIES OF
NITROGEN AND PHOSPHORUS FLOWS

From the preceding discussion two conclusions can be drawn. Many of the problems and questions raised are relatively new especially as they relate to general environmental considerations. Second, and perhaps as a consequence of the first, a very great deal is unknown. For example, much published work including references cited in this chapter is not always based on firm evidence. As a consequence there are difficulties in formulating strategies for a comprehensive investigation of the flows of nitrogen and phosphorus relevant to environmental problems.

A major environmental problem is the quality of surface and ground waters. Also, the hydrological cycle is the main link between various factors in the environment which eventually affect the quality of water. In the investigations described in following chapters, this simple idea provides the central theme for integrating the studies described.

At the outset, the watershed and its associated rivers and lakes was made the basic unit for study. The watershed contains all the factors relevant to the investigation, as represented in Figure 1.6.

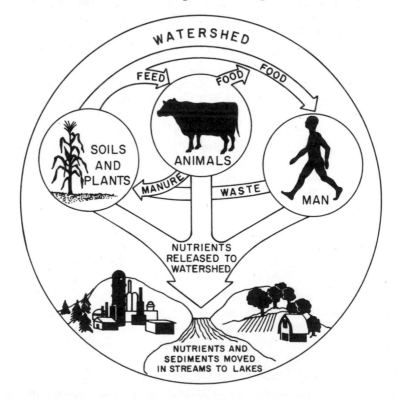

Figure 1.6. Nutrient flows within the human food web and from its components to the aquatic environment.

As water moves through the watershed, the forms and amounts of nutrients associated with it are affected by the soil, plants, animals and man himself, with corresponding consequences for the receiving waters. For the purposes of this study, it was decided to retrace the flows of water and nutrients from the final receiving water to their sources, in a sequence of steps as follows: (a) the lake, (b) the streams flowing to the lake, (c) the land drained by the streams, (d) the production of waste material which is eventually disposed on the land or into the stream, and (e) the social and economic factors affecting or affected by each step. The following chapters in general parallel this sequence corresponding to the specific objectives:

1. to quantify the relation between aquatic productivity and phosphorus transported into lakes;
2. to investigate the transport of sediment, phosphorus and nitrogen in streams and relate these to human activities;
3. to quantify the consequences of application of nutrients to land, considering especially the enrichment of ground water;
4. to evaluate the economic effects of nutrient management, especially on farm incomes;
5. to develop methods which control the level of nitrogen in wastes from animal production facilities prior to their disposal;
6. to identify social attitudes toward water quality, and to assess possible ways in which the public could directly respond to pollution problems; and
7. to propose guidelines for the management of nutrients in rural areas generated by human activity, and identify areas requiring further research.

The structure of this report parallels this outline.

Although the effects of the use of nitrogen and phosphorus in agriculture were a primary focus of this study, these effects should not be divorced from the consequences of man's broader activities. Where relevant, broader issues are discussed in the following chapters.

Where possible the use of the words eutrophication and pollution is avoided in subsequent discussions. Both terms are relative, and their use in some contexts may have inappropriate connotations. It is more accurate and less emotive to discuss the "enrichment" of water in terms of the levels of nutrients they contain and the measured consequences.

Figure 1.6, representing as it does the structure of the investigation, will be used as a symbol at the beginning of each chapter to indicate which aspect of the flow is being considered.

REFERENCES

Bigger, J. W., and R. B. Corey, 1970. Agricultural drainage and eutrophication. *In* Eutrophication: causes, consequences, correctives. Proceedings of a Symposium. National Academy of Sciences. Washington, D.C.

Cahn, R., 1972. Public perception of pollution control. *In* F. E. McJunken, ed. Costs of water pollution control. National Symposium. Water Resources Research Institute of the University of North Carolina.

Carson, R., 1964. Silent Spring. Houghton Mifflin Co. Boston.

Committee on Nitrate Accumulation, 1972. Accumulation of nitrate. National Academy of Sciences. Washington, D.C.

Council on Environmental Quality, 1970. First annual report. U.S. Government Printing Office. Washington, D.C.

————, 1971. Second annual report. U.S. Government Printing Office. Washington, D.C.

————, 1972. Third annual report. U.S. Government Printing Office. Washington, D.C.

————, 1973. Fourth annual report. U.S. Government Printing Office. Washington, D.C.

————, 1974. Fifth annual report. U.S. Government Printing Office. Washington, D.C.

The Environmental Pollution Panel, 1965. Restoring the quality of our environment. Report of the Environmental Pollution Panel. President's Science Advisory Committee. The White House. Washington, D.C.

Environmental Protection Agency, 1974. National water quality inventory. Washington, D.C.

————, 1973. Survey of facilities using land application of wastewater. R. H. Sullivan, M. M. Cohn and S. S. Baxter, eds. Prepared for Office of Water Program Operations. Washington, D.C.

Gamet, M. B., and J. M. Rademacher, 1960. Study of short filter runs with Lake Michigan water. J. Amer. Wat. Works. 52:2. pp. 137–152.

Hargett, N. L., 1974. 1974 Fertilizer summary data. National Fertilizer Development Center. Tennessee Valley Authority. Muscle Shoals, Alabama.

Hinton, W., 1966. Fanshen: A Documentary of Revolution in a Chinese Village. Random House. New York.

Hutchinson, G. E., 1957. A Treatise on Limnology. Volume 1. Geography, Physics, and Chemistry. John Wiley and Sons, Inc. New York.

Jacobs, J. J., W. H. Schaffer and G. L. Casler, 1973. An analysis of optimum use with minimum potential loss of nitrogen fertilizer on corn. J. Northeastern Ag. Ec. Council. 2:2. pp. 209–223.

Jewell, W. J., G. R. Morris, D. R. Price, W. W. Gunkel, D. W. Williams and R. C. Loehr. 1974. Methane generation from agricultural wastes: review of concept and future applications. Paper No. NA74–107. Amer. Soc. Ag. Engns. St. Joseph, Michigan.

Klein, L., 1962. River Pollution 11. Causes and Effects. Butterworths. London.

Loehr, R. C., 1974. Agricultural Waste Management, Problems, Process and Approaches. Academic Press. New York and London.

————, 1974. Characteristics and comparative magnitude of nonpoint sources. J. Wat. Pollut. Contr. Fed. 46:8, p. 1849.

National Academy of Sciences, 1969. Eutrophication: causes, consequences and correctives. Proceedings of a Symposium. Washington, D.C.

Odum, E. P., 1971. Fundamentals of Ecology. 3rd ed. W. B. Saunders Company. Philadelphia.

Russell, Sir E. John, 1971. The World of the Soil. Collins. London.

Sawyer, C. N., 1965. Problems of phosphorus in water supplies. J. Amer. Water Works Assoc. 65:11. p. 1431.

Stewart, B. A., 1970. A look at agricultural practices in relation to nitrate accumulation. *In* Nutrient Mobility in Soils: Accumulation and Losses. No. L. Special Publication Series Soil Science Society of America. Madison, Wisconsin.

Sullivan, R. H., M. M. Cohn and S. S. Baxter, 1974. Survey of facilities using land applications of wastewater. Prepared for Office of Water Program Operations. U.S. Environmental Protection Agency. EPA-430/a-73–006. Washington, D.C.

Task Group Report, 1967. Sources of nitrogen and phosphorus in water supplies. J. Amer. Water Works Assoc. 59:3. pp. 344–366.

———, 1966. Nutrient-associated problems in water quality and treatment. J. Amer. Wat. Works Ass. 58:10. pp. 1337–1355.

United States Department of Agriculture, 1973. Agricultural Statistics, 1973. U.S. Government Printing Office. Washington, D.C.

———, 1970. Number of feedlots by size groups and number of feed cattle marked. Statistical Reporting Service, Periodic Reports from 1966 to 1970. Washington, D.C.

———, 1955. U.S. Department of Agriculture Yearbook. U.S. Government Printing Office, Washington, D.C.

United States Department of Commerce, 1973. Statistical Abstract of the United States, 1973. U.S. Government Printing Office, Washington, D.C.

United States Environmental Protection Agency, 1974. National Water Quality Inventory 1974 Report to the Congress; EPA-440/9-74–001, Office of Water Planning and Standards, Washington, D.C.

Vallentyne, J. R., 1974. The algal bowl: lakes and man. Department of the Environment, Fisheries and Marine Service. Ottawa.

Principal Authors

K. S. Porter

J. Jacobs

D. A. Lauer

R. J. Young

2

The Response of Lakes to Phosphorus

Chapter 2

THE RESPONSE OF LAKES TO PHOSPHORUS

I will arise and go now, for always night and day
I hour lake water lapping with low sounds by the shore
While I stand on the roadway, or on the pavements grey,
I hear it in the deep heart's core.

—W. B. Yeats

INTRODUCTION: PHOSPHORUS AND AQUATIC PRODUCTIVITY

This chapter examines the algal standing crops of 13 lakes in central New York as related to inputs of phosphorus from the lake drainage systems. Considering the movement of phosphorus through a lake basin, this represents the final link in a chain of events that will be elaborated on in later chapters. Generalizations that may be inferred are also discussed here and in Chapter 8.

An increased supply of phosphorus to a temperate latitude lake causes a higher rate of primary production which, in turn, results in larger standing crops of algae. While this statement somewhat oversimplifies a complex ecological question, its general validity has been confirmed by a large body of research during the past several decades. Strong evidence that phosphorus is the nutrient limiting phytoplankton growth has been obtained for New York's Finger Lakes (Oglesby, *et al.*, 1975), a group which represents eleven of those studied.

Over a given period, the supply of phosphorus available to phytoplankton in a particular lake is a function of that entering the system from outside sources and that being cycled within the lake (Figure 2.1.). The latter has been the subject of considerable study by limnologists, but only in recent years have serious attempts been

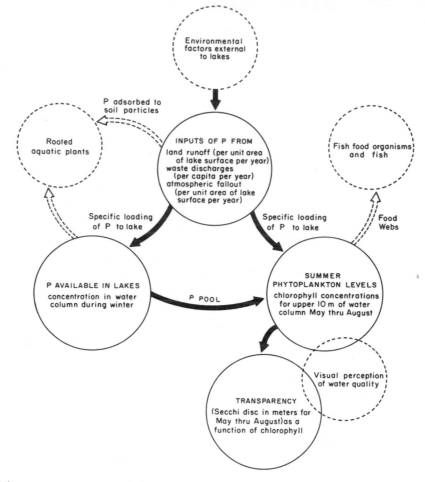

Figure 2.1. A conceptual diagram relating phosphorus (P) to primary productivity in lakes. Pathways are indicated by directional arrows and units of measure are given for the major components. Solid lines indicate aspects emphasized in this study.

made to estimate the transfer of phosphorus from watersheds to lakes. Internal recycling of phosphorus is likely to be relatively constant from year to year for a given lake and, for most lakes, is likely to be a relatively small fraction of that supplied annually from external sources. The processes involved are dependent upon factors not easily managed. On the other hand, changes in a lake's productivity over time and differences between lakes are largely functions of the phosphorus from outside sources, some of which may be controlled. Therefore, in order to develop strategies for optimizing the rate of phosphorus supply to lakes, attention must be directed to quantifying and comparing such sources. Costs and benefits, both qualitative and monetary, that would result from different control practices can be realistically ascertained only after this is done.

The productivity of lakes is limited during the winter by adverse environmental conditions, such as low levels of light and temperature, regardless of the availability of nutrients. More favorable growing conditions in spring and summer and the resulting algal production depletes available phosphorus and other nutrients in the water until one or more limits further growth. Algal cells thereafter decrease in numbers as they die and sink to the bottom or are eaten by herbivores. Increases will not occur until additional nutrients become available, from either internal recycling or external sources. A cyclic pattern of increase and decrease, highly variable from lake to lake or even for different years in the same lake, is thus established.

Because phytoplankton reproduce quickly relative to organisms that graze on them, favorable growth conditions can result in large standing crops of algae, known as "blooms." A high rate of nutrient supply, typically resulting in increased productivity beginning with phytoplankton and continuing through the aquatic food web, is called eutrophication.

Changes in the composition and relative abundance of species and in certain important chemical and physical parameters usually occur along with this increase in the overall biomass. Some changes may be undesirable, such as: (1) large standing crops of phytoplankton that interfere with some uses of water and decrease its transparency, (2) oxygen deficiencies occurring especially in lower depths of stratified lakes, and (3) changes in the balance between different types of fish and other organisms such that fishing is adversely affected.

OBJECTIVES AND ORGANIZATION OF LIMNOLOGICAL STUDIES

The main object of this study was to relate inputs of phosphorus to certain significant responses of lakes and to do so in a format that would be useful for management purposes. The responses chosen were summer phytoplankton standing crop and transparency of water. While both of these parameters are readily quantifiable, existing methods of estimating phosphorus input (loading) to a lake were judged inadequate. Therefore, a secondary objective became the development of a method for providing a more meaningful estimate of the phosphorus supplied to lakes from various external sources.

An initial approach was made toward developing a reasonably holistic model to simulate the response of lake phytoplankton and transparency to added phosphorus. The conviction that necessary data for this effort were neither available nor likely to become so during the course of the project caused this approach to be abandoned. This was fortuitous since it required the application of a more direct and simple approach of examining the relations between the parameters of concern without additional complicating factors. If successful, this simpler approach offered two other dividends: (1) the necessary data could be obtained with relative ease at a comparatively low cost and (2) the results could be applied to developing overall strategies for watershed management in terms that would be readily understood by the layperson.

Research was carried out to: (1) develop techniques for predicting phosphorus input to lakes as a function of land use, atmospheric fallout, and waste discharges; (2) relate levels of phosphorus in lakes to these inputs; (3) relate levels of phosphorus in lakes to phytoplankton standing crops; and (4) determine the effects of phytoplankton concentration (in terms of chlorophyll) on water transparency.

NATURE OF THE DATA

Production of phytoplankton is related to additions of phosphorus from land to lakes (Figure 2.1.). Detailed knowledge of this relation is essential if management guidelines for controlling phosphorus inputs are to be effective. Therefore, following others (e.g., Vollenweider, 1968; Patalas, 1972), detailed data obtained from the Fall Creek watershed (Chapter 3) were used to calculate phosphorus loss coefficients as a function of land use, as described below. These coefficients were then used to estimate the loss of phosphorus from land surfaces in other watersheds. This estimation required detailed information on area of land in various categories of use. Direct measurement of phosphorus concentrations and flows in a typical tributary stream and of phosphorus contributed by atmospheric fallout provided a uniquely well-defined data base for this study. The difficulties in obtaining such data for more than a very few watersheds are evident in Chapter 3.

The response of lakes to additions of phosphorus from external sources was gauged in terms of: (1) the concentration of phosphorus in the water column as measured in the winter; (2) summer standing crops of phytoplankton measured by chlorophyll concentrations in the upper 10 meters of the water column (May to August inclusive); and (3) water transparency during the summer, estimated from Secchi disc readings (see Figure 2.2.).

Figure 2.2. Secchi disc transparency being measured through the ice on Hemlock Lake. Secchi disc transparency is a measurement which indicates the visual clarity of water. To determine this, a target (Secchi disc) is lowered into the water on a calibrated line until it just disappears from view. The depth at which this occurs is the transparency and is analogous to the depth of water in which a shiny object, such as a beer can, would be just visible to a viewer peering down from a boat.

Once the inputs of phosphorus to a lake were defined, information on winter total phosphorus, summer phytoplankton, and transparency was needed for sequential periods. That is, the phosphorus added to a lake in a given year was presumed to be the primary determinant of total phosphorus in the water column during the following winter; this in turn served as an index of the phytoplankton standing crop the succeeding summer and transparency was simultaneously related to the latter. Enough reliable data of this nature could not be obtained from the literature. Therefore, ongoing studies of five of New York's Finger Lakes were modified slightly and expanded to other lakes to meet this need.

PHOSPHORUS INPUTS TO LAKES

Estimating Sources of Phosphorus

Principal sources of phosphorus contributed to lakes are usually domestic and, in some cases, industrial wastes, runoff from land, and fallout from the atmosphere. Subsurface inflows may supply large inputs of geochemically derived phosphorus to some lakes (e.g., Sylvester and Anderson, 1960). This can only be indirectly estimated by difference in an otherwise complete budget of inputs and outputs. In view of the low levels of phosphorus reported for ground waters in the Finger Lakes region, this is not believed to be an important phosphorus source for the lakes considered in this study.

Estimates of phosphorus inputs from atmospheric fallout were based on a program of measurement. Direct determination of this source may be important in assuring the accuracy of input calculations in view of the large regional differences reported in the literature (Pearson and Fisher, 1971; Owens, 1970; Weibel, 1969; Crisp, 1966; Chapin and Uttermark, 1973).

Wastes are discharged from municipalities, industries, and home disposal systems, usually septic tanks with tile drains. The phosphorus in municipal wastes on a per capita basis was estimated from published figures (Van Wazer, 1961; Engelbrecht and Morgan, 1959). Estimates of amounts discharged from sewage treatment plants were then calculated by adjusting for removal by various treatment processes. Data for industrial effluents were usually available.

The greatest difficulties occurred in the estimation of more diffuse sources of wastes for which direct measurement is virtually impossible. Oglesby, et al. (1973) have described an indirect method of estimating non-sewered domestic wastes, based on contributions of phosphorus to a stream from small unsewered villages located near a tributary of one of the Finger Lakes (Owasco). The only immediately practicable method of estimating phosphorus in land runoff was to measure or calculate that from point sources in a watershed, estimate that contributed by non-sewered households, and then to assume the remainder of phosphorus in streamflow was from land runoff. Extrapolation of such calculations to other watersheds requires careful consideration, however, given differences in soil type, slope, climate, and other factors.

The various parameters which we combined to estimate the phosphorus contributed by different sources could not all be defined for the same time period. For example, land use descriptions were for 1968 and population data came from the 1970 census. Inputs of phosphorus to lakes as calculated below thus represent a generalized pattern over a five-year period. Where appropriate, care was taken to allow for the effects of the New York State ban on phosphorus in household laundry detergents which became law in June of 1973.

Dynamics of Phosphorus in Lakes

Phosphorus may enter lakes in the form of ions of parts of dissolved molecules, crystals of various sizes, or as an adsorbate or inclusion of solid materials ranging in size from fine clay particles to gravel and smaller boulders. During periods of high runoff, the concentration of phosphorus associated with particulate material in suspension may be very high compared with that in solution. For example, Bouldin and Johnson (Chapter 3) found that under high flow conditions in Fall Creek the former ranged from 500 to 1,000 mg/m^3 while solubilized phosphorus concentrations were only 20 to 40 mg/m^3. Over the 20 months of their field study only about 23% of the phosphorus transported by Fall Creek was in dissolved form. What happens to the large quantity of particulate phosphorus when it enters a lake?

When water from a tributary flows into a lake, suspended material is subject to reduced turbulence and tends to settle to the bottom. This carries with it a large proportion of the associated phosphorus (Chapter 3). In addition, there is a net flux toward the bottom of that phosphorus which enters the lake in soluble form. Recent studies (Megard, 1971; Oglesby et al., 1973; Oglesby, MS 1974) show that up to 90% or more of the yearly phosphorus input is retained in large lakes. The only possible sink for this is the bottom sediment. Direct observation indicates that suspended matter carried into the Finger Lakes during high runoff periods settles to the bottom within days or weeks.

Species of Phosphorus

The previous generalizations about the probable fate of phosphorus associated with particulate matter introduced into lakes are an example not only of the importance of distinguishing between various states of this element but also of the difficulty in providing a functional description. Terminology describing various forms of phosphorus is an assortment of definitions and symbols confusing in its variety. This is due to the great number of chemical and physical combinations in which this element is found and to the fact that measurements based on available analytical methods have no biological meaning *per se*. Their significance must therefore be interpreted by the aquatic ecologist.

There are important differences between the forms of phosphorus in streams and in lakes. For example, in Fall Creek, a tributary of Cayuga Lake, most of the phosphorus transported annually is adsorbed to or included in suspended inorganic particles. In Cayuga Lake, phosphorus is predominantly a mixture of dissolved inorganic and organic forms plus that incorporated in living or dead organisms.

Disparate meanings have sometimes evolved for identical terms used by scientists working in different disciplines. Total phosphorus to a limnologist generally means all of that which is determined in a sample following moderately strong oxidation. This is a relatively simple and appropriate procedure for most lake waters but could markedly underestimate the amount in stream samples under high flow conditions when there are large amounts of particulate matter. Determination of total phosphorus in a stream sample should be by methods which include all the phosphorus in a sample, including the particulate forms, and sampling must include regimes of high flow (Chapter 3). There is at present no general agreement among aquatic scientists concerning what forms of phosphorus are biologically available and, in fact, there has been remarkably little discussion of this problem in the scientific literature.

Types of phosphorus referred to in this chapter are defined in Table 2.1. "Biologically available phosphorus" (BAP) is used here to describe inputs to lakes excluding the particulate phosphorus that both logic and the application of sorption isotherms (Chapter 3) indicated would not enter solution in the water column of a receiving lake. Forms of phosphorus relevant to the stream studies are described in Chapter 3.

Table 2.1. Terms used to describe the various forms of phosphorus (P) and phosphorus loading to lakes.

Symbol or term	Meaning	Use
MRP	Molybdate reactive phosphorus	That fraction in solution which reacts to form a molybdate blue complex following separation of suspended material by membrane (0.1–1 micron pore size) filtration or by centrifugation.
TP_L	Total phosphorus in lake water	Operational definition used for phosphorus in lakes that includes P in solution or suspension which reacts to form a molybdate blue complex following persulfate oxidation. This does not necessarily measure all P, for example occluded phosphorus, especially in stream samples.
Labile P	——	That amount of particulate phosphorus which sorption isotherms indicate is readily solubilized.
SUP	Soluble unreactive phosphorus	Phosphorus other than MRP in solution that reacts to form a molybdate blue complex following persulfate oxidation of sample previously membrane filtered or centrifuged.
BAP	Biologically available phosphorus (=MRP + SUP + labile P)	Operational definition used for the combination of those forms of P which would rationally appear to be available for biological uptake.
L_{sp}	Specific phosphorus loading (g TP/[m² lake surface · yr])	Refers to the loading of total phosphorus as defined by Vollenweider (1968).
L'_{sp}	Specific phosphorus loading (g "BAP"/[m² lake surface · yr])	Modification of L_{sp} with "biologically available phosphorus" substituted for total phosphorus.

Specific Phosphorus Loading

Vollenweider (1968), in an attempt to relate phosphorus in lakes to input in a biologically meaningful way, developed the concept of "specific phosphorus loading" (L_{sp}), the units of which are grams of phosphorus per unit area of lake surface per year. Specific loading is expressed on an areal rather than volumetric basis because the surface area of a lake is usually more representative of the euphotic zone than is the volume. This is especially so in the case of deep lakes where photosynthesis may occur in a relatively small fraction of the total volume. In shallow lakes, however, where there is more efficient recycling from the bottom sediments, areal loading may underestimate the annual supply of phosphorus. The rate of supply is considered on an annual basis because time lags that occur between input and biological uptake make finer resolution difficult to achieve.

Specific loading of biologically available phosphorus to the Finger Lakes (Figure 2.3) and several others in New York is considered to be the sum of the following: (1) phosphorus contributed from the atmosphere; (2) phosphorus from municipal sewage treatment plants; (3) phosphorus from non-sewered households; (4) runoff from residential areas; and (5) runoff from agricultural and forested land. These inputs of biologically available phosphorus were designated L'_{sp} to distinguish them from those of Vollenweider, which were based on total phosphorus.

Calculations for each respective input were as follows:

(1) Phosphorus in precipitation and other atmospheric fallout was from Likens (1972), who took measurements over a one-year period.

(2) Phosphorus from municipal sewage treatment plants was estimated from the number of people served by each plant multiplied by per capita annual contributions of phosphorus 1.5 kg/[cap · yr]) for the post World War II–1972 period (Engelbrecht and Morgan, 1959; Van Wazer, 1961; Hetling and Sykes, 1971) less removal by the plant (assumed to be 10% for primary, 20% for secondary, and 90% for tertiary treatment). Information on municipal sewage treatment plants was obtained from regional offices of the New York State Department of Environmental Conservation and also from operators of the facilities. Populations in the various lake basins were determined from 1970 census data (U.S. Bureau of the Census, 1970a and 1970b).

(3) Phosphorus from non-sewered households was as given for

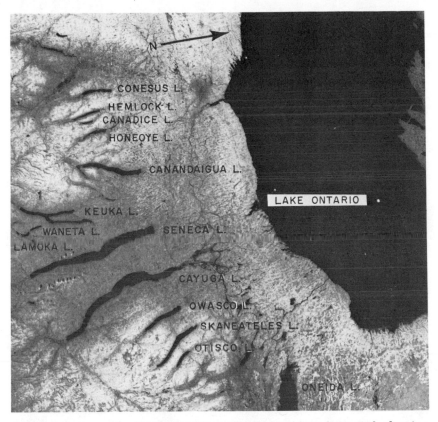

Figure 2.3. A satellite (NASA ERTS E-1234 15244-7 02) photograph showing most of the New York lakes considered in this study.

the Owasco Lake basin by Oglesby et al. (1973), who estimated this to be 50% of the annual per capita amount found in household wastes. This estimate appears high, but is explained by the tendency of populations in the northeastern U.S. to concentrate near streams and lakes.

(4) Phosphorus runoff from urban areas has been reported by a number of investigators (Weibel et al., 1964; Burm et al., 1968; Kluesener and Lee, 1972; Soderlund and Lechtinen, 1972; Sylvester, 1961) and is both high and quite variable from area to area. Because of the generally suburban nature of the residential areas in the Finger Lakes basins, a runoff coefficient of 100 mg/(m$^2 \cdot$ yr) was assumed for soluble phosphorus lost via urban runoff, which is low in the range of values reported for the storm drainage of cities.

(5) Corresponding estimates of runoff from agricultural and forest

land were obtained from the Fall Creek study (Chapter 3). In addition, well-defined land use data for all lake basins was available (Child, Oglesby, and Raymond, 1971) from New York's Land Use and Natural Resources Inventory (LUNR) of 1968 (Shelton et al., 1968).

To calculate L'_{sp}, stream transport of biologically available phosphorus was determined for the creek and its tributaries. The estimated amount of phosphorus from sewage (home disposal systems and treatment plants) was subtracted from the estimated total biologically available phosphorus to derive a fraction attributable to land runoff. Loss from forested land was calculated from data on several small watersheds where this was virtually the sole use. That from agricultural land was then determined by difference between these inputs and the total measured in Fall Creek. Runoff coefficients estimated by this method for agricultural and forested land were, respectively, 13.2 and 8.3 mg per m^2 land surface per year. In Chapter 3 a different method of apportioning the various factors contributing to phosphorus leaving the Fall Creek watershed was used, which may represent a more valid methodology, but introduces additional complications in extrapolating L'_{sp} calculations to other watersheds.

Calculated specific loadings (L'_{sp}) of biologically available phosphorus are given in Table 2.2, together with a few estimates of loadings for phosphorus in various forms based on recorded observations. The latter show poor agreement with calculated values for Canadarago and Oneida, but those for Cayuga and Owasco agree well.

Phosphorus loss coefficients for land runoff differ markedly when estimates of total phosphorus (Dillon and Kirchner, MS 1974) lost from agricultural land are compared with the values estimated for "biologically available phosphorus" in this study (Table 2.3). The differences in L_{sp} and L'_{sp} for the Finger Lakes are shown in Figure 2.4. Urban land runoff, considered together with the output of sewage phosphorus in the "human waste" category, accounted for an average of 11% (range = 5–15%) of the phosphorus.

PHOSPHORUS IN LAKES AND ITS
RELATION TO PHYTOPLANKTON STANDING CROPS

In the winter, lakes have higher levels of soluble phosphorus and lower concentrations of algae. Conversely, in the summer, algal numbers increase while concentrations of phosphorus diminish.

Table 2.2. Phosphorus loadings (g/[m² lake surface · yr]) for 13 lowland lakes in New York.

Calculated Loadings for 1972

Lake	L'_{sp}	Lake	L'_{sp}	Lake	L'_{sp}	Lake	L'_{sp}
Conesus	0.46	Honeoye	0.30	Seneca	0.53	Skaneateles	0.16
Hemlock	0.28	Canandaigua	0.28	Cayuga	0.67	Otisco	0.34
Canadice	0.22	Keuka	0.32	Owasco	0.66	Oneida	0.65
						Canadarago	0.80

Comparison of calculated loadings for different years and for loadings determined by other methods

Lake	L'_{sp}	Estimated loadings for various forms of phosphorus[1, 2, 3, 4]
Cayuga	0.67 (1972), 0.55 (1973–74) 0.45 (1974–75)	0.81, 0.54[1] (1970–71)
Owasco	0.66 (1972)	0.66[2] (1972)
Oneida	0.65 (1972), 0.44 (1974–75)	1.3[3] (1967–68)
Canadarago	0.80 (1972)	0.79, 0.44[4] (1969–70)

[1] For TP and MRP respectively. Data from Likens (1974) with direct sewage inputs corrected by Oglesby (MS 1974).
[2] For MRP. Data from Oglesby et al. (1973).
[3] For TP. Data from Greeson (1971).
[4] For TP and soluble P respectively. Data from Hetling and Sykes (1971).

These simple observations suggest inverse seasonal relations between nutrients and phytoplankton.

Sawyer (1947) used such relations to obtain crude predictions for summer algal levels based on winter phosphorus concentrations measured in a group of lakes in southern Wisconsin. He concluded

Table 2.3. A comparison of phosphorus loss coefficients (mg/[m² land surface · yr]) for runoff from land with different bedrock geologies and in different uses as determined from data collected in this study (Chapter 3) for "biologically available phosphorus" (BAP) and by Dillon and Kirchner (MS 1974) for total phosphorus (TP).[1] Ranges are given in parentheses.

Geology	Forest		Forest and Pasture		Agriculture	
	TP	BAP	TP	BAP	TP	BAP
Igneous	4.7 (0.7–8.8)	—	10.2 (5.9–16.0)	—	—	—
Sedimentary	11.7 (6.7–18.3)	8.3(?–?)	23.3 (11.1–37.0)	—	46 (approx.) (10–160)	13.2 (?–?)

[1] Mean obtained by averaging a large number of values from the literature.

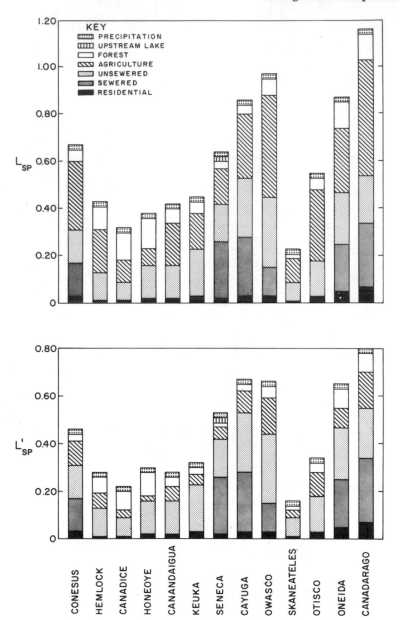

Figure 2.4. Specific phosphorus loadings for 13 New York lakes indicating inputs from sewage and residential runoff, agricultural lands, forests, and atmospheric fallout. Coefficients used in calculating land runoff were those of Dillon and Kirchner (MS 1974) for L_{sp} and those based on the data of Bouldin and Johnson (Chapter 3) in the case of L'_{sp}.

that those lakes with winter phosphorus concentrations greater than 10 mg/m³ (and nitrogen concentrations greater than 300 mg/m³) were likely to have undesirable growth of algae the following summer.

Sakamoto (1967) developed further the work of Sawyer by relating summer algal biomass in the epilimnion (expressed as the concentration of chlorophyll *a*, the primary photosynthetic pigment of green plants*) to winter or early spring phosphorus (expressed as the mean concentration of total phosphorus in the water column). Dillon and Rigler (1974) have subsequently added information on other lakes, particularly those in the Canadian Shield, to the extensive data for Japanese lakes cited by Sakamoto (1967). A log-log regression of summer chlorophyll on spring total phosphorus is shown in Figure 2.6. Winter total phosphorus is used here as an

Figure 2.5. Preparing a Van Dorn bottle for taking a water sample from Oneida Lake.

* Chlorophyll *a* concentration is a particularly useful index of the trophic state of a lake since it is related to transparency (Edmondson, 1972; Sakamoto, 1967) and can also be converted to indices of biomass, such as cell volume or particulate organic carbon (Strickland, 1960).

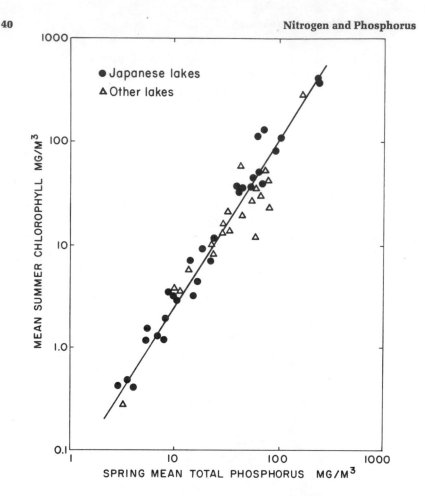

Figure 2.6. The relation between mean summer chlorophyll and spring total phosphorus concentrations for a number of Japanese and other lakes, from Dillon (MS 1974).

index, not an actual measurement, of the phosphorus eventually available for summer phytoplankton growth. For temperate latitude lakes that experience cold winters, most of this phosphorus should be in the soluble form since inputs of suspended material and phytoplankton production in the lake itself would be at an annual minimum when winter conditions have persisted.

The regression line in Figure 2.6 covers a wide range of values showing considerable scatter, especially at higher levels of total phosphorus and chlorophyll. From the data obtained for the lakes in New York State (Tables 2.4 and 2.5) a linear relation was obtained. The three lakes (4, 15 and 12c in Table 2.4) which appear to have higher chlorophyll concentrations per unit of winter phosphorus are all comparatively shallow (Table 2.4).

Table 2.4. Winter total phosphorus (TP) and mean depth ($\bar{Z}$) of some New York lakes.[1]

Lake	Lake No.	TP (mg/m³)	Standard Error of the Mean	Number of Samples	$\bar{Z}$ (m)
Conesus	1	17.6	3.7	6	11.5
Hemlock	2	10.9	0.8	6	13.6
Canadice	3	9.2		1	16.4
Honeoye	4	16.2		1	4.9
Canandaigua	5	10.1		1	38.8
Keuka 1972[2]	6a	8.2		1	30.5
Keuka	6b	15.1		1	30.5
Seneca[3]	7a	18.2		1	88.6
Seneca	7b	17.5		1	88.6
Cayuga 1972[4]	8a	20.7	2.4	5	54.5
Cayuga	8b	22.2		3	54.5
Cayuga 1974	8c	20.5		4	54.5
Owasco	9	14.7	0.5	5	29.3
Skaneateles	10	7.7	0.5	6	43.5
Otisco	11	8.4		1	10.2
Oneida 1968[5]	12a	28.9			6.8
Oneida 1969[5]	12b	29.0			6.8
Oneida	12c	42.0		1	6.8
Oneida 1975[6]	12d	31.1		2	6.8
Canadarago	13	39.5		1	6.7
Otsego	14	8.4		1	12.6
Waneta	15	24.0		1	3.5
Lamoka	16	13.9		1	5.0

[1] Samples collected in 1973 unless noted otherwise. Winter TP values represent the means of vertical profiles of samples taken on from 1 to 4 occasions during the winter.
[2] U.S. Environmental Protection Agency (1974a).
[3] U.S. Environmental Protection Agency (1974b).
[4] Data from Peterson and Barlow (Personal Communication).
[5] Data from Greeson (1971).
[6] Data from E. L. Mills (Personal Communication).

Regressions of summer chlorophyll on winter total phosphorus for New York lakes (Tables 2.4 and 2.5) are shown in Figures 2.7a–2.7c. The shallow lakes that do not exhibit a well-defined pattern of thermal stratification during the summer have been treated separately (Figure 2.7b) under the assumption that lakes which mix to the bottom during the warmer months probably have more effi-

Table 2.5. Summer chlorophyll measured in the epilimnion of New York lakes.[1]

Lake	Lake No.	Chlorophyll (mg/m^3)	Standard Error of the Mean	Number of Samples
Conesus 1971	1a	8.8		
Conesus 1972	1b	4.5	0.7	13
Conesus	1c	5.6	0.7	16
Hemlock 1971	2a	6.8		3
Hemlock 1972	2b	6.7	0.7	13
Hemlock	2c	5.8	0.4	16
Canadice	3	4.4		
Honeoye	4	13.2		
Canandaigua	5	2.6		
Keuka	6	3.3		
Seneca	7	7.1		
Cayuga 1972	8a	9.7	0.9	13
Cayuga	8b	7.8	0.6	12
Cayuga 1974	8c	8.7	0.7	15
Owasco 1971	9a	6.4	1.0	6
Owasco 1972	9b	4.2	0.8	13
Owasco	9c	6.2	0.5	18
Skaneateles 1971	10a	1.0		
Skaneateles 1972	10b	1.8	0.3	13
Skaneateles	10c	1.5	0.1	16
Otisco	11	2.2		
Oneida	12	29.0		
Canadarago	13	22.5		
Otsego	14	2.2		
Waneta	15	20.8		
Lamoka	16	7.3		

[1] Samples collected in 1973 unless noted otherwise. Each sample represents an average of concentrations measured at 5-m intervals in the water column with additional values added for the 2-m depth during 1973.

cient recycling of phosphorus as organic detritus is introduced into the food web found in the water column. This should be reflected in larger concentrations of summer chlorophyll per unit of winter total phosphorus for the shallow lakes, as is true for the New York lakes considered in this study. A comparison of the intercepts of the regression lines in Figures 2.7b and 2.7c also suggests that phosphorus may be used more efficiently in shallow lakes. The high correlation for the above regressions gives some confidence in the relations. This is especially encouraging because it suggests that only a relatively small amount of data on winter total phosphorus is needed for a particular lake to predict mean summer chlorophyll concentration.

The lack of information in the literature which was needed to define the various relations has already been noted. However, data on Lake Washington for summer chlorophyll and winter total phosphorus was one noteworthy exception. This was not referred to in developing the regression shown in Figure 2.7, but rather reserved as a test to see how this regression might hold up for a deep lake located in another part of the United States. Lake Washington is a uniquely valuable test case because measurements cover a period when sewage inputs varied widely (Edmondson, 1972). Over the period 1957–63, inputs of phosphorus from domestic sewage were high and increasing, until 1963 when the input decreased as the sewage was diverted, finally ceasing to enter the lake after February 1968.

Figure 2.8 depicts the changes in summer chlorophyll as a function of winter total phosphorus for Lake Washington compared to the regression equation for deep lakes in New York State. When the pattern of change is considered as a whole, it is apparent that this equation would have been an extremely good predictor of the course of events in Lake Washington.

SPECIFIC LOADING AND WINTER
TOTAL PHOSPHORUS CONCENTRATIONS

If, as suggested previously, winter total phosphorus is an index of the phosphorus in a lake available to support growth of phytoplankton, it is equally reasonable to suppose that winter total phosphorus is a function of specific loading. Such proves to be the case and a regression of winter phosphorus on loading is shown in Figure 2.9. Oneida and Canadarago, both shallow and with high specific loadings, appear to deviate from the relation in a manner that supports the above idea that phosphorus is more efficiently recycled in such lakes. The correlation coefficient is again high enough and the 95% confidence intervals sufficiently narrow to promote confidence in the regression.

SUMMER CHLOROPHYLL AND
SPECIFIC PHOSPHORUS LOADING

From the close correlations between summer chlorophyll and winter total phosphorus and between the latter and specific loading, correlation between summer chlorophyll and phosphorus loading to a lake would be anticipated. This is verified in Figure 2.10. Points are more scattered than for the previously discussed relations, but the pattern is still sufficiently well-defined to indicate the nature of the dependence, with the shallower lakes again fitting the model least satisfactorily.

The regression of summer chlorophyll on specific loading is especially important since it represents a direct link between phosphorus inputs and a water quality parameter of considerable concern to the public. Unlike the use of winter total phosphorus as an index of summer phytoplankton standing crop, it is a relation readily understandable to the non-scientist and hence easily adapted as part of any overall model that might be developed for water quality management.

CHLOROPHYLL AND TRANSPARENCY

The standing crop of algae and water transparency when considered together provide the maximum quantifiable information available to describe the aesthetic properties of water (see Chapter 7 for a further discussion of visual perception). Transparency, as measured by the Secchi disc, is primarily a function of the concentration of fine suspended particles in the water column, but may be influenced by colloidal organic materials, by the reflectivity of light on the surface of the water, or by the size and reflectivity

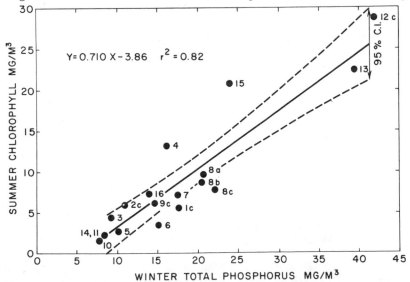

Figure 2.7a–c. Relation between mean summer chlorophyll in the epilimnion and winter total phosphorus for: (a) all lakes in Tables 2.4 and 2.5, (b) the shallow lakes, and (c) the deep lakes.

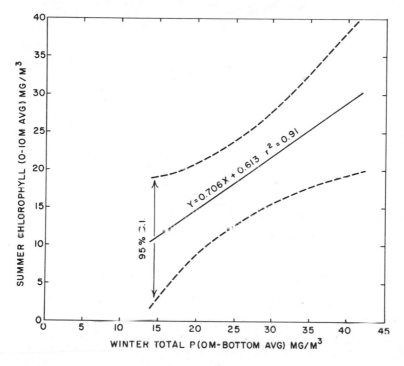

b.

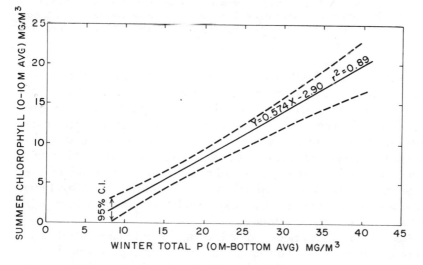

c.

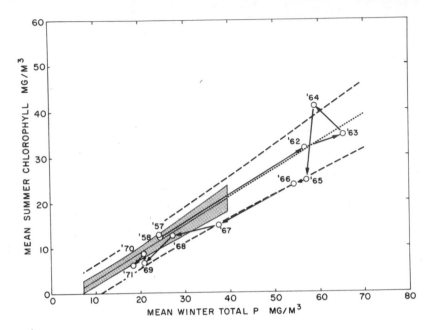

Figure 2.8. Changes in summer chlorophyll concentrations in Lake Washington as a function of the winter total phosphorus (TPL) concentration during periods of changing phosphorus loading to the lake. The regression from Figure 2.7c (shaded portion) is extended (dotted line) to encompass the range of values measured in Lake Washington. The arrows are years. Confidence intervals (95%) for predicted individual mean summer chlorophyll concentrations are within the dashed lines. The hatched area designates the 95% confidence intervals about the mean as determined for the deep New York lakes.

of the Secchi disc itself. Soil particles and detritus as well as cells or colonies of planktonic organisms decrease the transparency of the water. For this reason, to correlate chlorophyll concentrations (taken as representative of the phytoplankton cells in suspension) with Secchi disc readings, samples containing large quantities of inorganic suspended particles should be excluded. Data taken from the New York lakes after periods of heavy rainfall, especially during those months when vegetative cover was minimal, were excluded, as were those from shallow lakes such as Oneida, where wind stirred up bottom sediments.

Figure 2.11 relates Secchi disc transparency to concentrations of chlorophyll. About 350 observations, representing 80% of the total available data (those omitted having abnormally low transparencies for the chlorophyll that was present), were used in the calculations.

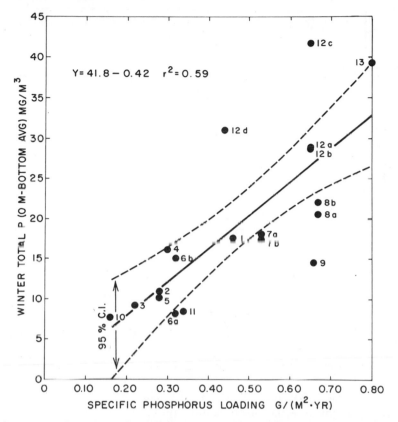

$$Y = 41.8 - 0.42 \quad r^2 = 0.59$$

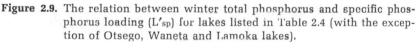

Figure 2.9. The relation between winter total phosphorus and specific phosphorus loading (L'_{sp}) for lakes listed in Table 2.4 (with the exception of Otsego, Waneta and Lamoka lakes).

Additional, unscreened data for Lake Washington (Edmondson, 1972) was included in order to extend the range of the curve to higher values of chlorophyll. It is evident from Figure 2.11 that as chlorophyll concentrations exceed 15–20 mg/m^3, there is no perceptible difference in transparency. As first pointed out by Edmondson (1972), substantial reductions of phytoplankton standing crops in highly eutrophied lakes may not result in significant increases in transparency.

DISCUSSION AND SUMMARY
Review of Other Studies

In recent years there have been attempts to mathematically describe the production of standing crops of phytoplankton in lakes.

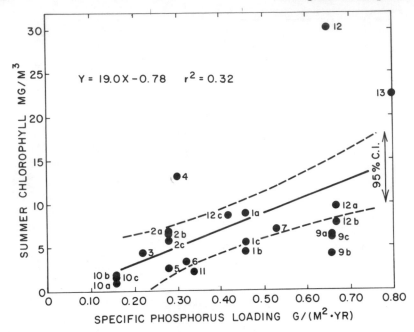

Figure 2.10. The relation between summer chlorophyll (mean for the epilimnion) and specific phosphorus loading for those lakes listed in Table 2.3 with the exception of Otsego, Waneta and Lamoka.

Most of these have been efforts to describe growth in relation to nutrients from external sources and nutrient cycling within lakes, and have been either purely theoretical (e.g., Patten, 1973) or have described single lakes (e.g., Larson et al., 1973). Shannon and Brezonik (1972), using multivariate analysis to isolate factors associated with the trophic status of a number of Florida lakes, obtained evidence that human population and land use around these lakes were highly correlated with eutrophication.

Probably the most complete model for a natural system is that of DiToro et al. (1971) for the phytoplankton population in the Sacramento–San Joaquin Delta. Equations expressing the kinetics of mass transport (nutrients, phytoplankton, and growth) were combined with other equations describing processes of biotic change. Rate coefficients were estimated from the literature and are probably among the weakest features of the model. Initial tests against data obtained for the San Joaquin River gave encouraging results, but the model's usefulness for other systems is questionable since some parameters of the model were constructed using information specific to the Sacramento–San Joaquin Delta system.

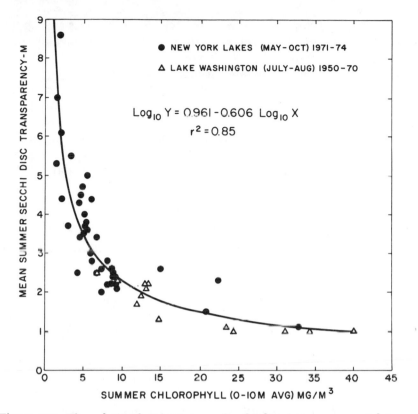

Figure 2.11. The relation between mean Secchi disc transparency and mean summer epilimnetic chlorophyll for thirteen New York lakes and Lake Washington.

Vollenweider (1968), in a study of the relationship of phosphorus input and several physical parameters of lakes to trophic state, employed a different method using extant data and deducing key variables. He related trophic state to annual total phosphorus loading and to either mean lake depth or to the mean depth divided by water retention time (Vollenweider and Dillon, 1974). A large number of lakes throughout the world form crude trophic groupings under this treatment. The use of these "models" to develop management strategies is hampered by the imprecisely defined "trophic state" and by the inability to predict quantified changes in phytoplankton standing crops in response to changes in phosphorus loading. Two simple but important aspects of Vollenweider's study were the selection of a year as the basic unit of time and the expression of phosphorus loading on the basis of amount per unit surface area.

One of the authors of this chapter (Oglesby, 1969) was among the early advocates of a relatively holistic modeling approach using sets of differential equations to describe the biotic and associated abiotic variables which determine net primary production in a lake. This idea was explored during the first year of the project and finally rejected due to the lack of reliable rate coefficients for growth kinetics of natural phytoplankton communities and the virtual impossibility of representing mixing processes in lakes. If models representing an advance over those of Vollenweider were to be developed in this study, restriction to the use of actual data within rational, simple frameworks was necessary.

Discussion of Limnological Studies

During the second two years of this project, using data largely from lakes located in upstate New York, relations between phosphorus loading, the phosphorus annually available to lake biota (as indexed by lake winter total phosphorus), the summer phytoplankton levels in lakes, and water transparency were defined. Interdependencies of these variables are summarized in Figure 2.12, where specific phosphorus loading (L'_{sp}) is annual and chlorophyll and transparency represent summer means.

Calculations of specific loading in terms of "biologically available phosphorus" (BAP) indicate a much lower fraction coming from agricultural sources than would have been the case had total phosphorus (TP) been used as a measure of phosphorus loading. Agricultural sources of phosphorus appear to be less important for the New York lakes studied than reported by other investigators who used TP as a measure of loading for other lakes. Total phosphorus measured in streams includes at least a portion (dependent upon the analytical methods employed) of the phosphorus adsorbed to soil particles that originated from such sources as agricultural land runoff.

Changes in specific loading that could be anticipated if various management practices were adopted are shown in Table 2.6. For example, if all agriculture in the Cayuga Lake basin were eliminated and the land allowed to revert totally to forest, the L'_{sp} would decrease by only 4%. However, the 1973 New York ban on phosphate in household laundry detergents is calculated to have reduced loading by 37%; and if additional treatment for phosphorus removal were added to existing sewage treatment plants, the total decrease would be 58%. A resultant L'_{sp} of 0.28 g/(m² lake surface · yr) would

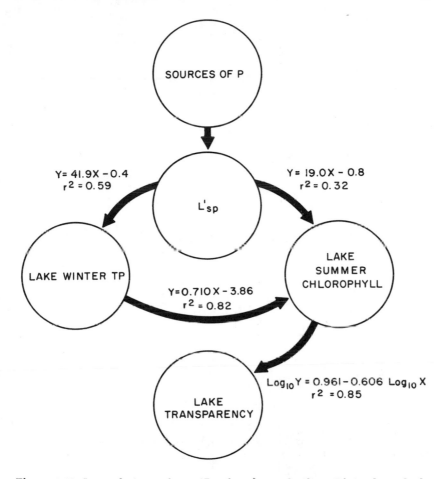

Figure 2.12. Interrelations of specific phosphorus loading (L'sp), phytoplankton standing crop, and transparency for a group of New York lakes.

be estimated (Figures 2.10 and 2.11) to give a mean chlorophyll concentration of 4.2 mg/m^2 and an average Secchi disc transparency of 3.75 m during the summer.

Transparency, as measured by the Secchi disc, is related to the summer standing crop of phytoplankton in the lakes studied (Figure 2.11). Since improvement of the water's appearance is frequently an aim of lake management, it is significant that for very high concentrations of phytoplankton, reductions in their concentration produce little, if any, change in transparency.

Extensive data on summer chlorophyll and winter total phos-

Table 2.6. Changes in calculated specific loading (L'_{sp}) (g/[m² lake surface · yr]) and % reduction associated with various phosphorus management strategies.

Lake	1970–72	A If all agriculture converted to forest		B With present ban on detergent P		C If 90% P removed at sewage treat. plants		B + C	
	L'_{sp}	L'_{sp}	% red.	L'_{sp}	% red.	L'_{sp}	% red.	L'_{sp}	% red.
Conesus	0.46	0.42	9	0.31	33	0.34	26	0.26	44
Hemlock	0.28	0.26	7	0.22	21	—	—	0.22	21
Canadice	0.22	0.21	4	0.18	18	—	—	0.18	18
Honeoye	0.30	0.29	3	0.23	23	—	—	0.23	23
Canandaigua	0.28	0.25	11	0.21	21			0.21	21
Keuka	0.32	0.31	3	0.22	31			0.22	31
Seneca	0.53	0.51	4	0.32	40	0.30	43	0.22	58
Cayuga	0.67	0.64	4	0.42	37	0.45	33	0.28	58
Owasco	0.66	0.61	8	0.45	32	0.56	15	0.35	39
Skaneateles	0.16	0.15	6	0.12	25	—	—	0.12	25
Otisco	0.34	0.30	12	0.26	24			0.26	24
Oneida	0.65	0.62	5	0.44	32	0.48	26	0.35	46
Canandarago	0.80	0.74	8	0.56	30	0.56	30	0.45	44

Figure 2.13. Fishermen on the shore of Cayuga Lake.

phorus were available for only five of the New York lakes included in this study, but limited data from other lakes were also useful in developing cause and effect hypotheses. The summer chlorophyll–winter total phosphorus relationship appears to be valid for other lakes in temperate latitudes, although limited evidence for shallow lakes with short water retention times suggests these may have phosphorus–phytoplankton relationships similar in nature, but quantitatively different from those of deep lakes.

Through the North American Lakes Project, testing of all the relations of specific loading, winter total phosphorus, summer chlorophyll, and transparency should be possible in the near future.

Recommendations for Further Research

It was beyond the scope of this project to relate production of phytoplankton to that of fish and their food organisms. However, in developing programs of lake management, this is an important factor that does not presently receive enough attention. Simple estimations indicate that fish production could be expected to vary in direct proportion to changes in phytoplankton production.

Another significant problem can be the growth of rooted aquatic plants. These may use phosphorus adsorbed to soil particles and can also act as a pool of labile phosphorus. It follows that lakes with extensive beds of littoral plants may deviate from the results obtained in this study.

ADDENDUM

Subsequent to the writing of this chapter, a significant advance has been made in how the phosphorus loading to lakes is calculated and its relation to winter total phosphorus and mean summer chlorophyll concentrations. In brief, loading has been redefined as the annual input of phosphorus to the summer mixed zone (epilimnion) of a lake or to the volume encompassed by its mean depth, whichever is least, rather than to the surface area. This is rationalized as being more representative of phosphorus input to the euphotic zone where the phytoplankton are being produced. One effect of expressing loading in this manner is to positively weight the phosphorus supplied to shallow, wind-mixed lakes relative to those that are deep and thermally stratified.

Thus, with surface area loadings the predicted responses (summer standing crops of phytoplankton or winter total phosphorus concentrations) are the same regardless of lake depth and patterns of vertical mixing. Using the new method of calculating loading, the shallower the mixed zone of a lake the greater would be the predicted response. Arithmetically this is a consequence of multiplying surface area by the depth of the thermocline (or mean depth of the lake if this is less) in the loading calculation with the effect of diluting phosphorus added to lakes with deep epilimnia, e.g., Cayuga, relative to those with shallower mixed depths, e.g., Conesus, or that are unstratified, e.g., Honeoye. The true cause and effect relationships involved are incompletely understood at this time but probably include a more efficient recycling of phosphorus in shallow lakes due to resuspension of bottom sediments in response to wind mixing.

The regressions obtained using loading to the mixed zone represent a considerable improvement over those calculated on an areal basis. They are as follows:

$$\hat{Y} = 0.23X - 3.36 \qquad r^2 = 0.73 \qquad (1)$$

where,

$\hat{Y}$ is the mean summer concentration of chlorophyll a plus phaeopigment in the upper 10 m of the water column

X is the annual loading of BAP to the mixed zone or to the volume encompassed by the mean depth contour, whichever is least

$$\hat{Y} = 0.338X + 0.992 \qquad r^2 = 0.83 \qquad (2)$$

where,

$\hat{Y}$ is the winter concentration of total phosphorus in the water column

X is the same as in (1) above

Details of both methods and results will be submitted for publication in professional journals in the near future.

REFERENCES

Burm, R. J., D. F. Krawezk and G. L. Harlow. 1968. Chemical and physical comparison of combined and separate sewage discharge. Jour. Water Pollution Contr. Fed. 40:112–126.

Chapin, J. D. and P. D. Uttormark. 1973. Atmospheric contributions of nitrogen and phosphorus. Report of the Water Resources Center, University of Wisconsin, Madison. 35 p.

Child, D., R. T. Oglesby and L. S. Raymond, Jr. 1971. Land use data for the Finger Lakes region of New York State. Cornell Univ. Water Resources and Marine Sciences Center, Ithaca, N.Y. Publ. No. 33. 29 p.

Crisp, D. T. 1966. Input and output of minerals for an area of pennine moorland: the importance of precipitation, drainage, peat erosion and animals. Jour. Appl. Ecol. 3:327–348.

Dillon, P. J. MS 1974. Progress report on the application of the phosphorus loading concept to eutrophication research. A report prepared on behalf of R. A. Vollenweider for NRC Associate Committee on Scientific Criteria for Environmental Quality Subcommittee on Water, Canada Centre for Inland Waters, Burlington, Ont.

Dillon, P. J. and W. B. Kirchner. MS 1974. The effects of geology and land use on the export of phosphorus from watersheds.

Dillon, P. J. and F. H. Rigler. 1974. The phosphorus-chlorophyll relation in lakes. Limnol. Oceanogr. 19(5):767–773.

DiToro, D. M., D. J. O'Connor and R. V. Thomann. 1971. A dynamic model of the phytoplankton population in the Sacramento-San Joaquin Delta. Adv. in Chem. No. 106. p. 131–180.

Edmondson, W. T. 1972. Nutrients and phytoplankton in Lake Washington. In Nutrients and eutrophication: the limiting-nutrient controversy, G. E. Likens, ed. American Society of Limnology and Oceanography. Spec. Symposia, 1:172–188.

Engelbrecht, R. S. and J. J. Morgan. 1959. Studies on the occurrence and degradation of condensed phosphates in surface waters. Sewage Ind. Wastes. 31:458–478.

Greeson, P. E. 1971. Limnology of Oneida Lake with emphasis on factors contributing to algal blooms. U.S. Geological Survey open-file report. 185 p.

Hetling, L. J. and R. M. Sykes. 1971. Sources of nutrients in Canadarago Lake. New York State Department of Environmental Conservation, Tech. Paper No. 3. Albany. 35 p.

Kluesener, J. W. and G. F. Lee. 1972. Nutrient loading from a separate storm sewer in Madison, Wisconsin. Paper presented at 45th Water Pollution Control Federation Conference, Atlanta, 1972.

Larson, D. P., H. T. Mercier and K. W. Malueg. 1973. Modeling algal growth dynamics in Shagawa Lake, Minnesota, with comments concerning projected restoration of the lake. *In* E. J. Middlebrooks, D. H. Falkenborg and T. E. Maloney, eds. Modeling the eutrophication process. Proceedings of Workshop, Utah State Univ., Logan, Utah, Sept. 5–7, 1973.

Likens, G. E. 1972. The chemistry of precipitation in the central Finger Lakes region. Cornell Univ. Water Resources and Marine Sciences Center, Ithaca, N.Y. Tech. Rept. 50. 47 p.

Likens, G. E. 1974. Water and nutrient budgets for Cayuga Lake, New York. Cornell Univ. Water Resources and Marine Sciences Center, Ithaca, N.Y. Tech. Rept. 82. 91 p.

Megard, R. O. 1971. Eutrophication and the phosphorus balance of lakes. Paper presented to American Society of Agricultural Engineers, Chicago, Ill., Dec. 7–10, 1971.

Mills, E. L. 1975. Personal communication.

Oglesby, R. T. 1969. Effects of controlled nutrient dilution on the eutrophication of a lake. *In* Eutrophication: causes, consequences, correctives. Proceedings of International Symposium on Eutrophication, Madison, Wisc., June 11–15, 1967. National Academy of Sciences, Washington, D.C. p. 483–493.

Oglesby, R. T. MS 1974. The limnology of Cayuga Lake, New York: Monograph prepared for the North American Lakes Project.

Oglesby, R. T., L. S. Hamilton, E. L. Mills and P. Willing. 1973. Owasco Lake and its watershed. Cornell Univ. Water Resources and Marine Sciences Center, Ithaca, N.Y. Tech. Rept. 70. 52 p. + appendices.

Oglesby, R. T., W. R. Schaffner and E. L. Mills. 1975. Nitrogen, phosphorus and eutrophication in the Finger Lakes. Cornell Univ. Water Resources and Marine Sciences Center, Ithaca, N.Y. Tech. Rept. 94. 27 p.

Owens, M. 1970. Nutrient balances in rivers. Water Treat. Exam. 19:239–252.

Patalas, K. 1972. Crustacean plankton and the eutrophication of St. Lawrence Great Lakes. Jour. Fish. Res. Board Canada. 29:1451–1462.

Patten, B. C. 1973. Need for an ecosystem perspective in eutrophication modeling. *In* E. J. Middlebrooks, D. H. Falkenborg and T. E. Maloney, eds. Modeling the eutrophication process. Proceedings of Workshop, Utah State Univ., Logan, Utah, Sept. 5–7, 1973

Pearson, F. J., Jr. and D. W. Fisher. 1971. Chemical composition of atmospheric precipitation in the northeastern United States. U.S. Geological Survey Water-Supply Paper 1535P. 23 p.

Peterson, B. L. and J. P. Barlow. 1972. Personal communication.

Sakamoto, M. 1967. Primary production by phytoplankton community in some Japanese lakes and its dependence on lake depth. Arch. Hydrobiol. 62(1):1–28.

Sawyer, C. N. 1947. Fertilization of lakes by agricultural and urban drainage. Jour. New England Waterworks Assoc. 61:109–127.

Shannon, E. E. and P. L. Brezonik. 1972. Eutrophication analysis: a multivariate approach. Proc. Amer. Soc. Civil Engineers. SA1:37–57.

Shelton, R. L., E. E. Hardy and C. P. Mead. 1968. Classification, New York State land use and natural resources inventory. Center for Aerial Photographic Studies, Cornell Univ., Ithaca, N.Y. Mimeo.

Soderland, G. and H. Lechtinen. 1972. Comparison of discharges from urban stormwater runoff, mixed storm overflow, and treated sewage. Paper presented at the 16th International Water Pollution Control Federation Conference, Atlanta, 1972.

Strickland, J. D. H. 1960. Measuring the production of marine phytoplankton. Jour. Fish. Res. Board Canada. 122:1–172.

Sylvester, R. O. and G. Anderson. 1960. An engineering and ecological study for the rehabilitation of Green Lake. Report to the Seattle, Washington Park Board. 194 p.

Sylvester, R. O. 1961. Nutrient content of drainage water from forested, urban and agricultural areas. *In* Algae and metropolitan wastes. R. A. Taft Sanitary Engineering Center, Cincinnati, Ohio. p. 80–87.

U.S. Bureau of the Census. 1970a. Census of population. General population characteristics. Final Report PC(1)–B34, New York. U.S. Gov't. Printing Office, Washington, D.C.

———. 1970b. Census of housing. Block statistics for selected areas of New York State. Final Report HC(3)–163. U.S. Gov't. Printing Office, Washington, D.C.

U.S. Environmental Protection Agency. 1974a. Report on Keuka Lake. Nat'l. Environmental Research Center, Las Vegas, Nev. Working Paper No. 160. 29 p.

———. 1974b. Report on Seneca Lake. Nat'l. Environmental Research Center, Las Vegas, Nev. Working Paper No. 170. 57 p.

Van Wazer, J. R. 1961. Phosphorus and its compounds. Interscience, New York.

Vollenweider, R. A. 1968. The scientific basis of land and stream eutrophication, with particular reference to phosphorus and nitrogen as factors in eutrophication. Tech. Rpt. OECD, Paris, Directorate for Scientific Affairs/CSI/68. 27:1–182.

Vollenweider, R. A. and P. J. Dillon. 1974. The application of the phosphorus loading concept to eutrophication research. National Research Council of Canada. Publ. No. 13690. 42 p.

Weible, S. R. 1969. Urban drainage as a factor in eutrophication. *In* Eutrophication: causes, consequences, correctives. Proceedings of International Symposium on Eutrophication, Madison, Wisc., June 11–15, 1967. National Academy of Sciences, Washington, D.C. p. 384–408.

Weibel, S. R., R. J. Anderson and R. L. Woodward. 1964. Urban land runoff as a factor in stream pollution. Jour. Water Pollution Contr. Fed. 36:914–924.

ACKNOWLEDGMENTS

Edward L. Mills, Paul J. Godfrey and Kenneth H. Pollock, all former or present Cornell graduate students, were responsible for collecting and processing most of the data pertaining to the Finger Lakes with assistance from research technicians James McKenna and Myra Ginsparg. Additional funding for the Finger Lakes' studies was provided by the Office of Water Resource Technology, U.S. Department of the Interior (matching grants A-047-NY and B-038-NY) and, in the case of Owasco Lake, by Cayuga County.

Principal Authors

Ray T. Oglesby
William R. Schaffner

3

The Influence of Human Activity on the Export of Phosphorus and Nitrate from Fall Creek

Chapter 3

TRANSPORT IN STREAMS

In the space of 176 years the Lower Mississippi has shortened itself 242 miles. That is an average of a trifle over 1-1/3 miles a year. Therefore, any calm person, who is not blind or idiotic, can see that in the Old Oolitic Silurian Period, just a million years ago next November, the lower Mississippi River was upward of 1,300,000 miles long, and stuck out over the Gulf of Mexico like a fishing rod. And by the same token any person can see that 742 years from now the Lower Mississippi will be only 1-3/4 miles long, and Cairo and New Orleans will have joined their streets together. . . . There is something fascinating about science. One gets such wholesome returns of conjecture out of such a trifling investment of fact.

—Mark Twain, *Life on the Mississippi*

The objective of this chapter is to discuss the influence of human activities on the phosphorus and nitrate removed from the Fall Creek watershed in the stream discharge.

GENERAL DESCRIPTION OF THE WATERSHED

The Fall Creek watershed lies in the cool-temperature zone of the U.S. with a humid, continental climate. Storms are cyclonic, bringing moisture from the Gulf of Mexico and the Atlantic Ocean. Average annual precipitation ranges from about 85 cm (34 in) near Ithaca to about 100 cm (40 in) in the eastern-most portions of the watershed. The precipitation is fairly uniformly distributed throughout the year. Approximately 50% of the precipitation is lost from the watershed as streamflow, with most of the discharge during the winter and spring months. Elevations range from about 600 m (1800 ft) in the northern and eastern parts to 380 m (830 ft) at the gaging station at Forest Home. At the higher elevations the soils are derived from acid or low-lime glacial till while at the lower elevations and along streams the soils are derived from a complex mixture of glacial outwash and lacustrine and alluvial deposits. The unconsolidated mantle is underlain at variable depths by gently dipping siltstones and shales. At the Forest Home gaging station the stream flows through a small gorge cut in the bedrock.

Total area is 330 km². Land use is summarized in Table 3.1. The major agricultural enterprise is dairy farming with corn and hay production primarily utilized as feed for the dairy cows. There are 6800 cows on about 130 dairy farms. Most of the dairy farms are found in the Virgil Creek subwatershed and in the Fall Creek sub-watershed above Freeville.

Table 3.1. Summary of land use for the Fall Creek watershed estimated from a 1968 inventory (Shelton *et al.* 1968).

Land Use Category	Total Hectares	Approximate % of Total Area
High intensity cropland	95	.3
Cropland	12,124	36.0
Permanent pasture	1,683	5.0
Inactive agriculture	5,977	17.7
Forest land and plantations	11,448	34.0
Shrub wetland	745	2.1
Wooded wetland	229	.7
High and medium density urban	334	1.0
Low density urban	64	.2
Strip development	69	.2
Rural hamlet	153	.5
Commercial strip development	87	.3
Public and semi-public	627	1.9
Ponds, lakes and streams	49	.1
	33,684	100.0

Total population is about 12,000 people. Except for the secondary sewage treatment plant serving about 1400 people in Dryden, sewage disposal is primarily by individual septic tanks and associated disposal fields. Some of the disposal fields do not function properly in certain portions of the watershed either because of seasonal high water tables or relatively impermeable substrates. The effluent from these poorly functioning fields often drains into road ditches and natural drainage ways. Several mobile home parks and apartment complexes dispose of sewage through lagoons that discharge into natural drainage ways.

The watershed is typical in many respects of much of the southern two or three tiers of counties in New York and of portions of northern Pennsylvania. The bedrock, soils, and land use are similar to those of much of this region. Precipitation ranges from 65 cm (30 in.) to 110 cm (44 in.) per year over the region, with corresponding variability in average annual stream discharges.

In the more northern portions of western New York, the nature of the soils, the unconsolidated mantle, and the bedrock are different from those in southern New York. The landscape slopes are generally less steep and the fraction of the land under intensive cultivation increases. The precipitation is generally lower, the population densities are generally higher. Thus the Fall Creek watershed is not typical of these areas and the results have an uncertain application to these and other landscapes.

GENERAL PROCEDURES

The primary sampling site was about 100 meters upstream from the U.S. Geological Survey gaging station at Forest Home (location 1, Figure 3.1), which is about 6 km upstream from the point where Fall Creek empties into Cayuga Lake. Discharge records at this station were available for the period 1926 through the present. The daily flow duration hydrograph and the cumulative discharge and suspended solids loading is illustrated in Figure 3.2. Evidently, a large fraction of the water and even more of the suspended solids leave the watershed during a small fraction of the time. Accordingly, sampling schemes were carried out with the objective of defining the composition of the water during periods of both high and low discharge rate. In addition, the changes in composition of the water with changing discharge rate provide valuable clues about the sources of phosphorus and nitrate, as will be illustrated later. The samples were taken on an irregular schedule with up to 12 samples per day during periods of high discharge rate and at intervals of 1 to 3 weeks during periods of low discharge rate.

At the Forest Home site, the stream flows over a low dam and through a narrow bedrock channel which maintain high turbulence. Samples collected within a short time of each other at different locations across the stream were very similar in composition which indicates the stream was well mixed at this point.

Through cooperative arrangements with the U.S. Geological Survey, gaging stations were established on two streams draining subwatersheds (locations 6 and 16, Figure 3.1), a staff gage was installed and calibrated on Fall Creek at Freeville (location 15, Figure 3.1), and flow measurements were made at selected sites occasionally.

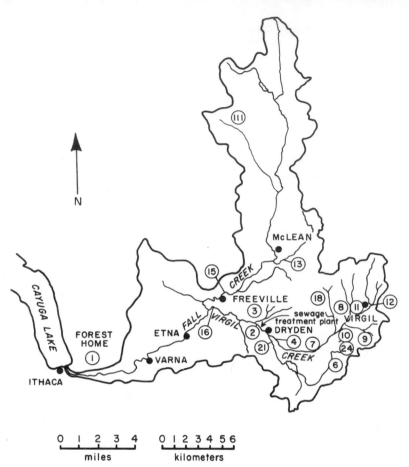

Figure 3.1. Outline map of Fall Creek basin showing location of major subwatershed sampling stations.

Samples were taken for chemical analysis at approximately 20
locations throughout the watershed on an irregular basis. A portion
of these are shown by identifying site numbers on Figure 3.1.

PHOSPHORUS
Rationale of Procedures

The phosphorus carried in streams is distributed among several
chemical forms and physical states. Consideration of the nature and
behavior of these several components suggests that their effect on
the biology of lakes differs, although definitive data on which to

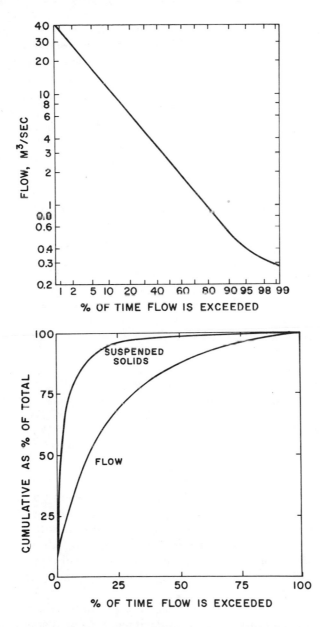

Figure 3.2. (Upper) Flow duration hydrograph of Fall Creek at the Forest Home gaging station. (Lower) Cumulative amount of suspended solids and flow plotted against percent of time the flow is exceeded.

base quantitative interpretations are not available. Some important considerations in making judgments about the relative biological effects are summarized in the following paragraphs. A major fraction of the total phosphorus carried out of a watershed by a stream is associated with the particulate matter suspended in the water, much of which is delivered during a relatively short period of time when the discharge rate is very high and hence the velocity of flow is sufficient to transport particles of varying sizes and densities. A smaller fraction of the total phosphorus is dissolved in the water. For example, during storm periods Fall Creek water often contained 500 to 1000 micrograms of phosphorus per liter associated with particulate matter and 20 to 40 micrograms per liter of dissolved phosphorus. Upon reaching the lake these fractions behave very differently; the larger particles soon settle to the bottom because water movement in the lake is no longer sufficient to keep them in suspension. At least in the lakes in central New York, a few days after a large storm which delivers large amounts of suspended solids, the water in the lake (even near stream outlets) becomes relatively clear and only the fine particles remain in suspension plus, of course, the dissolved constituents. Lake samples seldom contained more than a total of 10 to 50 micrograms of phosphorus per liter in all forms. Thus the phosphorus remaining in the lake water is very different in nature and amount from the phosphorus delivered by the stream during storm periods.

The above observations lead to the question of whether or not 1 kg of particulate phosphorus will have the same effect on the biology of lakes as 1 kg of dissolved phosphorus. According to the reasoning which follows, the amount of particulate phosphorus has only a minor effect on the biology of lakes, while the dissolved phosphorus has a major effect.

First, the bottom of the lake is covered with sediment delivered by past events from the lake basin. If the phosphorus in the sediment delivered by past events was essentially in the same forms as that delivered by current events, accumulation of more of the same seems unlikely to change the effect of the bottom sediments on lake biology; that is, adding a few millimeters of sediment to several meters of sediment seems unlikely to have an important effect on lake biology.

Second, questions can be raised about the chemical activity of the particulate phosphorus; for example, if samples of suspended solids were leached with phosphorus-free water, how rapidly would the solid phase phosphorus dissolve and what would be the concentration of phosphorus in the leaching water? The results of such studies show that there is a small amount (usually considerably less than 10%) of the total phosphorus which dissolves fairly rapidly and that the concentration of leachate decreases as leaching continues. Such phosphorus is referred to as labile or sorbed phosphorus. Once this labile phosphorus is removed, the remainder dissolves very slowly and concentrations in solution are only a few micrograms per liter. Furthermore, this is consistent with the fact most of the phosphorus associated with the suspended solids has persisted through long geological periods; its resistance to dissolution must be considerable or it would have been leached out of the unconsolidated mantle and carried to the sea long ago. The results of studies with Fall Creek sediments demonstrate that the particulate matter contains about 1.1 milligrams of total phosphorus per gram and about 0.04 milligrams (4%) of this is in a labile or sorbed form.

Separation of dissolved or soluble phosphorus from particulate phosphorus is an arbitrary procedure, particularly when the objective is to produce fractions differing markedly in biological effects. Careful examination of stream samples should reveal a continuum of particle sizes from coarse sand to molecules and ions. The term "dissolved" would properly include only ions and molecules. However, some of the very fine particulate matter may contain phosphorus which is solubilized during the determination of phosphorus by the molybdate procedure (which requires acidic conditions). Thus there is no exact procedure for separating "dissolved" from "soluble" phosphorus.

Another perplexing question arises with respect to phosphorus which is a component part of organic molecules or very finely divided organic matter. Many biologically mediated reactions convert organic forms of phosphorus to biologically active forms. However, some phosphorus in organic forms may be resistant to conversion to biologically active forms so even here uncertainty prevails.

Most of the questions discussed above will not be resolved in a completely satisfactory manner for a long time. Considering these uncertainties, the investigators of this project devised an operational fractionation procedure for phosphorus in stream samples and developed a set of working hypotheses about the influence these operationally defined fractions of phosphorus have on the biology of lakes. The success of this procedure has already been demonstrated in the previous chapter where variability among lakes was consistent with the expectations based on the working hypotheses; in other words, the approximations seem to be useful in describing the influence of stream inputs on the biology of lakes. Our present understanding of an array of complex processes does not, however, allow us to be more definite than to state "consistent with expectations based on working hypotheses."

Several fractions of stream phosphorus were defined on the basis of operational procedures. The details of the fractionation and analytical procedures are described in Appendix A. The first fractionation was an approximate separation of dissolved phosphorus and particulate phosphorus. To perform this separation, the samples were centrifuged at a relative centrifugal force of 35,000 times gravity for 30 minutes. Filtration was not practical because the large amount of suspended solids in some samples clogged the filters rapidly. The centrifugation removed mineral particles larger than 0.02×10^{-3} mm in diameter, but less dense organic particles of larger size remained suspended. The phosphorus in the supernatant was further subdivided on the basis of reactivity with molybdate reagent; an aliquot of the supernatant was treated with molybdate reagent and the phosphomolybdate complex was extracted into isobutanol, and reduced, and the intensity of the resulting blue color was determined. The results of this analysis were termed molybdate reactive phosphorus and abbreviated MRP. This fraction includes the dissolved inorganic phosphorus, the phosphorus associated with the small particles which is dissolved by the very acid molybdate reagent, and organic phosphorus which is hydrolized during the analytical procedure. The results of studies not reported here indicate that most of the MRP is in fact dissolved inorganic phosphorus.

A second fraction of phosphorus in the supernatant was determined by the molybdate procedure following treatment of an aliquot of the supernatant with persulfate. The persulfate oxidation presumably converts all (or most) of the phosphorus in organic combinations to inorganic form which then reacts with the molybdate reagent. The results of this latter procedure we refer to as "total soluble phosphorus" and abbreviate TSP. A third fraction in the supernatant can be defined on the basis of the difference between TSP and MRP; this fraction we refer to as "soluble unreactive phosphorus" or SUP.

The particulate matter from selected samples was analyzed for total phosphorus. The amount of sorbed or labile phosphorus was determined by a sorption isotherm method.

Thus the analysis of the stream samples yielded essentially four fractions of phosphorus which potentially differ in biological effects. Based on the preceding discussion the following hypotheses were formulated about the effects of these fractions. Probably 100% of the molybdate reactive phosphorus (MRP) and most of the total soluble phosphorus (TSP) is available to organisms in lakes. In the remainder of this chapter most emphasis will be placed on this fraction because of the evident success of this fraction in explaining differences among lakes as documented in the preceding chapter.

Probably the labile or sorbed fraction of particulate phosphorus is of limited importance to the biology of lakes and, as will be documented later, in Fall Creek the amount of such phosphorus is small. Probably the remainder of the particulate phosphorus is of no consequence to the biology of lakes, but, for the sake of completeness, the suspended solids and phosphorus content of the suspended solids was determined. Thus even though major emphasis is on TSP, the total particulate phosphorus was determined and it is reported here even though not much emphasis was placed on the results in interpreting human influence on lakes.

Particulate Phosphorus

The loading of particulate phosphorus was calculated as the product of suspended solids and the average fraction of phosphorus in suspended solids. In following sections, the amount of suspended solids carried past Forest Home and its phosphorus content is discussed.

Loading of Suspended Solids at Forest Home

Two hundred ml aliquots of samples, taken as described previously, were centrifuged at a relative centrifical force of 1500 × gravity for 30 minutes. The supernatant was decanted, and the remaining sediment was dried at 105°C and weighed.

The suspended solids concentration in samples taken at the Forest Home site is plotted against discharge rate in Figure 3.3 for the period November 1972 through April 1973. These data are representative of the range and variability of the data collected throughout the observation period. For a given discharge rate, the concentration of suspended solids is higher if the discharge rate is increasing (rising limb of storm hydrograph) than if it is decreasing (receding limb). To facilitate calculation of loading, empirical regression equations of the form:

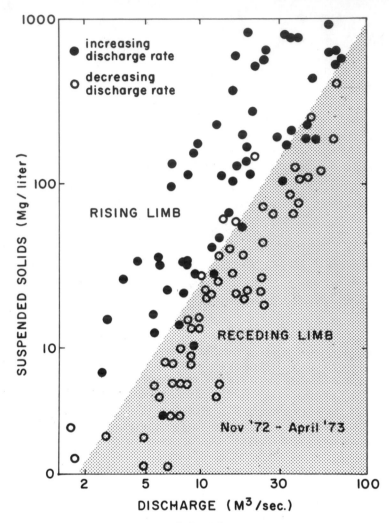

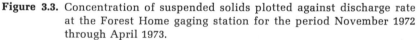

Figure 3.3. Concentration of suspended solids plotted against discharge rate at the Forest Home gaging station for the period November 1972 through April 1973.

$$C_{ss} = a_0 + a_1Q + a_2\frac{\Delta Q}{\Delta t} \qquad (3.1)$$

where,

C_{ss} = concentration of suspended solids (mg/1)
Q = discharge rate (m³/sec)
$\dfrac{\Delta Q}{\Delta t}$ = rate of change of discharge rate (m³/sec/hr)

a_0, a_1, a_2 are regression coefficients

were calculated by months for the two high runoff periods (November 72–April 73, December 73–April 74). Parameters of the regression equations are given in Table 3.2.

Table 3.2. Parameters for regressions of concentration of suspended solids (mg/l) on discharge rate (m³/sec) and rate of change of discharge rate (m³/sec/hr). Equation 3.1. n is the number of observations, R is the multiple correlation coefficient.

Month/Year	a_0	a_1	a_2	n	R
November 1972	—81.4	7.9	63.4	26	.94
December 1972	—50.0	6.1	35.5	23	.97
January–February 1973	—59.5	13.2	87.1	18	.96
March 1973	—90.7	15.0	75.4	21	.87
April 1973	—22.2	4.3	28.0	18	.92
November 1972 to					
April 1973	—17.8	6.1	64.0	106	.86
December 1973	—20.3	11.0	69.2	23	.74
January–February 1974	—52.1	13.1	143.0	16	.99
March 1974	—34.5	6.7	31.0	16	.90
April 1974	—15.3	4.7	72.7	35	.80
December 1973 to					
April 1974	—14.9	6.2	117.2	90	.79
November 1972 to					
April 1974	— 9.7	6.3	71.5	196	.81
June 21–26, 1972	—44.3	5.4	48.9	14	.98

Loading was calculated for each month during the experimental period by summing loadings calculated for each 2-hour period. Discharge rate data was generally reported by the U.S. Geological Survey at intervals of 2 hours, and the average discharge rate over 2-hour periods was defined as the reported bi-hourly discharge rate.

During intervals of high discharge rate, the concentration of suspended solids was calculated from the appropriate regression equations and corresponding loading was calculated for the intervals of high discharge rate according to Equation 3.2.

$$\Sigma SS = K\Sigma(Q)\left(a_0 + a_1 Q + a_2 \frac{\Delta Q}{\Delta t}\right) \qquad (3.2)$$

where,

ΣSS = loading of suspended solids in metric tons
Q = discharge rate (M³/sec)
$\dfrac{\Delta Q}{\Delta t}$ = rate of change of discharge rate (m³/sec/hr)

a_0, a_1, a_2 are regression constants listed in Table 3.2

$K = 7.2 \times 10^{-3}$ which converts the product of discharge rate and concentration of suspended solids to metric tons/2 hours.

For intervals of low discharge rate, the loading of suspended solids was calculated as the product of accumulated discharge and average concentration.

The results of these calculations are listed in Table 3.3 for the experimental period.

Table 3.3. Calculated loading of suspended solids at Forest Home for the period September 1972 through April 1974.

Month	Year	Discharge, $m^3 \times 10^6$	Loading of Suspended Solids, metric tons	Standard Error
Sept.–Oct.	1972	6.8	214	±51
November	1972	27.7	2,941	±336
December	1972	40.9	4,288	±934
January[1]	1973	20.6	1,912	±357
February[1]	1973	16.6	2,716	±242
March	1973	29.9	3,815	±828
April	1973	32.3	2,159	±213
May	1973	18.0	567	±136
June	1973	8.2	258	±62
July	1973	3.2	101	±24
August	1973	2.1	66	±16
September	1973	2.4	76	±18
October	1973	2.1	66	±16
November	1973	3.2	101	±24
December	1973	21.6	2,987	±893
January[2]	1974	18.4	1,607	±158
February[2]	1974	15.4	1,610	±132
March	1974	21.0	755	±224
April	1974	39.0	4,365	±420
TOTAL		330.3	29,155	±1722

[1] January and February loading is calculated from January–February '73 regression (Table 5.1).

[2] January and February loading is calculated from January–February '74 regression (Table 5.1).

During this investigation considerable effort was expended in trying to derive reasonable estimates of the reliability of the calculated loadings. This question seems so important yet is so often not even mentioned that the following discussion is presented with the objective of illustrating the problems. Two general problems are discussed: a) the reliability of the estimates of the loadings calculated for the periods covered by the study and b) the reliability of estimates based on extrapolation of the experimental data to other time periods.

The standard errors of the loading estimated for high discharge rate intervals are listed in Table 3.3 and were derived from the variance calculated using Equation 3.3:

$$\text{Var}\,(\Sigma SS) = (K\Sigma Q)^2\,\text{Var}\,(a_0) + (K\Sigma Q^2)^2\text{Var}(a_1)$$
$$+ \left(K\Sigma Q\,\frac{\Delta Q}{\Delta t}\right)^2\text{Var}(a_2) + 2(K\Sigma Q)(K\Sigma Q^2)\text{Cov}(a_0 a_1)$$
$$+ 2(K\Sigma Q)\left(K\Sigma Q\,\frac{\Delta Q}{\Delta t}\right)\text{Cov}(a_0 a_2)$$
$$+ 2(K\Sigma Q^2)\left(K\Sigma Q\,\frac{\Delta Q}{\Delta t}\right)\text{Cov}(u_1 u_2)$$

(3.3)

For low discharge rate intervals, the standard errors were calculated as the product of accumulated flow and standard error of the average concentration of suspended solids.

In a stream the size of Fall Creek all of the suspended solids exported cannot be collected and measured directly; therefore, interpolation procedures, such as those outlined here, must be used. With such procedures, three sources of error can be recognized: a) those associated with sampling and measuring the concentration of suspended solids in the samples, b) those associated with interpolation between sampling intervals and c) those associated with estimating discharge. The standard errors listed in Table 3.3 are an estimate of (a) and (b); however, any bias in sampling (such as not sampling bed load) is not included. Thus the standard errors listed are primarily deviations from regression, which is composed of lack of fit of the model and random variation associated with sampling and sample analysis. Since replicates of discharge measurements are not available, the uncertainties in discharge are not included in the standard errors, but these uncertainties are probably relatively small at the Forest Home site. (The control section at the stage height recorder is stable since it is a concrete dam set in the gorge walls and the relationship between stage height and discharge is based on about 600 flow measurements made over 50 years.)

Two procedures were used to generalize the results and estimate an average yearly export of suspended solids based on 30 years of discharge records. The first estimate was derived by calculating the regression of observed suspended solids loading per month on observed discharge for that month. This regression was then used to estimate monthly loading from average monthly discharge observed over the last 30 years. The regression is illustrated in Figure 3.4 and the expected average loading for each month is listed in Table 3.4. This method yields an average suspended solids loading of about 14,000 metric tons per year, or 42 T/KM^2.

A second estimate based on the method of Archer and LaSala (1968) yielded an average annual loading of 14,350 metric tons.

Two aspects of reliability of estimates for nonexperimental periods can be considered: a) those associated with estimates for a specific month or year and b) those associated with average loading over a long period of time (e.g., 30 years). With respect to a specific

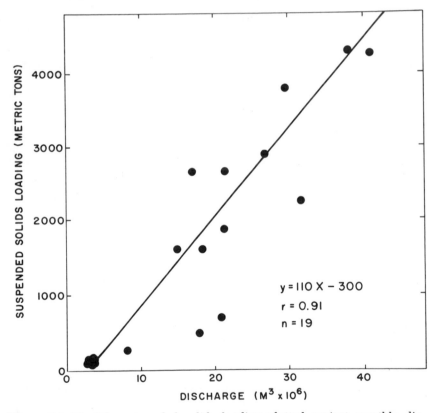

Figure 3.4. Monthly suspended solids loading plotted against monthly discharge for the period November 1972 through April 1974. Points are experimental observations, the line represents the regression equation.

Table 3.4. Estimates of average monthly loading of suspended solids calculated from the regression in Figure 3.4.

Month	Mean Discharge[1] $m^3 \times 10^6$	Suspended Solids Loading, metric tons
January	14.2	$1{,}262 \pm 696$[3]
February	14.7	$1{,}317 \pm 698$
March	32.8	$3{,}308 \pm 785$
April	29.4	$2{,}934 \pm 765$
May	17.0	$1{,}570 \pm 706$
June	10.7	877 ± 687
July	4.7	217 ± 677
August	3.0	30 ± 675
September	3.3	63 ± 675
October	2.7	0 ± 675
November	11.7	987 ± 689
December	13.7	$1{,}429 \pm 701$
TOTAL	159.9	$13{,}992$[2] $\pm 2{,}436$[4]

[1] Mean discharge 1943–1973.

[2] This is equivalent to 42 metric tons/km^2/yr or 120 tons/mi^2/yr.

[3] $S_y = S_y._x \left(1 + \dfrac{1}{n} + \dfrac{X^2}{\Sigma X^2} \right)^{\frac{1}{2}}$

[4] $(\Sigma S_y{}^2)^{\frac{1}{2}}$

time interval (a specific year or month) the uncertainties seem so large as to make any estimate of loading meaningless. For example, in the period 1943 to 1973, total discharge per year varied from $80 \times 10^6 m^3$ to $250 \times 10^6 m^3$. Examination of Table 3.3 and Figure 3.4 illustrates considerable variability in loading of suspended solids between a given month for the 2 years of the study and variability among months with similar discharge. Use of the average flow for a 30-year period (as was done to derive loading shown in Table 3.4) should produce a fairly reliable estimate of the average loading expected over a 30-year period, providing a reasonable sampling of the population of climatic events was obtained during the experimental period. Because of the relatively short period of measurement, the array of characteristics of climatic events may not be typical of those likely to be encountered over 30 years. This may bias in a major way both the estimated mean and the estimated variance of the population of yearly loading. Thus the statistical analysis presented here is biased to the extent that the characteristics of the climatic events during the experimental period were not representative of the population of climatic events.

Particulate Phosphorus in Fall Creek

The total phosphorus content of about 200 samples of suspended solids averaged 0.11% phosphorus (standard error of mean, 0.03% P). The estimated average loading of particulate phosphorus was calculated as the product of expected suspended solids loading (14,000 metric tons) times 0.0011. This yields about 15 metric tons of phosphorus per year for the whole watershed, or 46 kg P/km² or 0.46 kg P/ha or 46 mg P/m². The standard error of these estimates is about 25% of the estimate.

Comparison with Other Watersheds

Illustrated in Figure 3.5 are frequency distributions for suspended solids loading for 29 watersheds in western New York (Archer and LaSala, 1968; Gilbert, 1967). The values reported for Fall Creek are lower than about 2/3 of the 29 watersheds. As another reference point, Menard (1961) calculates the geological erosion rate in the Appalachian region is now about 20 metric tons/km²/yr. Varioni *et al.* (1970) summarized the loading for the continental U.S. by regions and the Fall Creek values are low compared to most of these values.

Assuming that the total phosphorus content of suspended solids is 0.11% in the western New York area, the range in particulate

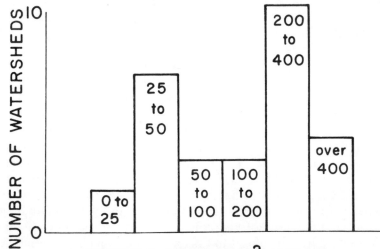

Figure 3.5. Distribution of annual loading of suspended solids for 29 western New York watersheds (Archer and LaSala, 1968; Gilbert, 1967).

Figure 3.6. Canoeing on Fall Creek. Note eroding stream bank in the background.

phosphorus loading would be from about 28 mg P/m^2/yr for 25 metric tons of suspended solids per km^2/yr to 440 mg P/m^2/yr for 400 metric tons of suspended solids per km^2/yr. Thus, the loading of particulate phosphorus probably varies tremendously among watersheds because of the large differences in suspended solids loading.

Sorbed Phosphorus

Over the last several years, soil scientists have demonstrated that there is a labile fraction of soil phosphorus which can be usefully described by sorption isotherm procedures. Taylor and Kunishi (1971) have utilized this procedure for studies of suspended solids in streams, and in a conceptual way others have utilized this idea (Ryden, Syers and Harris, 1973). Furthermore, a sodium bicarbonate extractant is generally considered to extract an amount of phosphorus from soils which is proportional to this labile or sorbed fraction.

The labile phosphorus content of selected samples of suspended solids was determined using the sorption isotherm method of Taylor and Kunishi (1971). Determinations on fresh samples indicate that 40 micrograms phosphorus per gram suspended solids is a reasonable estimate. This indicates that about 600 kg of labile phosphorus is carried on the suspended solids past Forest Home per year.

Total Soluble Phosphorus

In the following sections the data are analyzed with the objective of estimating the amount of molybdate reactive phosphorus (MRP) and total soluble phosphorus (TSP) which the various sources contributed to the stream discharge.

Results at Forest Home

Figure 3.7 illustrates the concentrations of MRP at the Forest Home sampling site plotted against discharge rate for a portion of the experimental period. Loading of MRP at Forest Home was calculated using procedures similar to those for suspended solids. Regressions of MRP on discharge rate and rate of change of discharge rate were calculated by months and the regression parameters are listed in Table 3.5.

The MRP loading was calculated as the sum of bi-hourly loading using Equation 3.2 and appropriate parameters from Table 3.5. For those months not shown in Table 3.5, loading was calculated as the product of total flow and average concentration. Variance was cal-

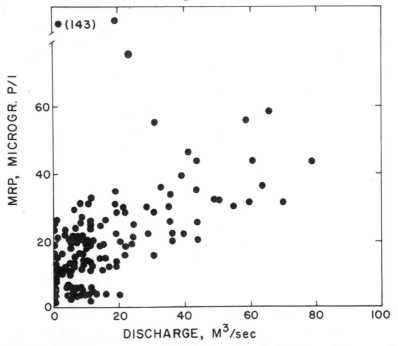

Figure 3.7. The MRP concentrations in samples at the Forest Home gaging station plotted against discharge rate for the period September 1972 to August 1973.

culated as described for suspended solids. The results for the experimental period are listed in Table 3.6, together with standard errors and weighted mean concentration (total MRP loading/total discharge).

Table 3.5. Parameters[1] for regression of MRP on discharge rate and rate of change of discharge rate at the Forest Home gaging station.

Month	Year	b_0	b_1	b_2	n	R
November	1972	16.6	.56	1.68	28	.91
December	1972	16.4	.27	.42	15	.91
February	1973	22.7	.21	.48	9	.73
March	1973	13.8	.16	.65	26	.34
April	1973	6.0	.50	3.92	17	.82
November– April	1972 to 1973	13.2	.41	1.30	95	.72
December	1973	9.8	.88	1.58	81	.69
January	1974	12.4	.42	.21	39	.46
February	1974	13.2	.43	1.93	38	.81
March	1974	9.6	—.06	.27	43	.07
April	1974	8.6	.15	.33	85	.54
December– April	1973 to 1974	13.1	0.13	0.86	286	.26
November– April	1972 to 1974	12.3	0.26	1.38	381	.48

[1] $MRP = b_0 + b_1 Q + b_2 \frac{\Delta Q}{\Delta t}$, MRP is micrograms P per liter, Q is discharge rate in m^3/sec, $\frac{\Delta Q}{\Delta t}$ is rate of change of discharge rate in $m^3/sec/hr$.

There were no statistically significant correlations of SUP with discharge either during high discharge rate intervals or low discharge rate intervals. The mean was 12 microg P/l (standard deviation of population 11.7, standard error of mean 0.6). Accordingly, the loading of SUP was calculated as the product of mean concentration (12 microg P/l) and total discharge with standard errors calculated as previously described for suspended solids. The results for the experimental period are listed in Table 3.6.

With these results as a base, attention is focused in the next section on allocating this total soluble phosphorus to various sources and human activities in the watershed.

Estimates of Biogeochemical TSP

The phrase "biogeochemical TSP" is used to mean the TSP which would be delivered by the watershed if there were no human activity in the watershed.

Table 3.6. Loading of soluble phosphorus for the experimental period September 1972 to April 1974 at Forest Home.

Month	Year	Discharge $m^3 \times 10^6$	MRP Loading, kg P	MRP S.E.[1]	MRP Weighted Mean Conc. µg P/l	SUP Loading kg P	TSP Loading kg P
Sept-Oct	1972	6.8	80	11	12	19	99
November	1972	27.7	775	29	28	332	1107
December	1972	40.9	944	38	23	491	1435
January	1973	20.6	517	38	25	247	764
February	1973	16.6	432	27	26	199	631
March	1973	29.9	482	33	16	359	841
April	1973	32.3	523	46	16	388	911
May	1973	18.0	117	18	7	216	333
June	1973	8.2	205	81	25	98	303
July	1973	3.2	29	6	9	38	67
August	1973	**2.1**	20	**3**	**10**	25	45
September	1973	2.4	24	2	10	29	53
October	1973	2.1	13	2	6	25	38
November	1973	3.2	24	2	8	38	62
December	1973	21.6	493	20	23	259	752
January	1974	18.4	305	14	17	221	526
February	1974	15.4	279	13	18	185	464
March	1974	21.0	188	18	9	252	440
April	1974	39.9	500	16	13	479	979
TOTAL		330.3	5950± 127[2]		18	3900± 170[3]	9850 ±210[4]

[1] Standard error calculated as described in Equation 3.3.

[2] $[\Sigma(S.E.)^2]^{\frac{1}{2}}$

[3] $[(\Sigma Q)^2 Var(\text{mean concentration of SUP})]^{\frac{1}{2}}$

[4] $[Var(MRP) + Var(SUP)]^{\frac{1}{2}}$

Estimates of biogeochemical TSP were derived from measurements of water samples taken from streams draining subwatersheds 24 and 111 (see Figure 3.1), in which there were no homes or farms and from wells not adjacent to barns, feedlots, homes, etc. Watershed 24 is covered mostly with trees of various ages and with some brush. It has been cut over and perhaps farmed at some time, but its exact history is unknown. The soils are derived from glacial till and the pH of the surface horizons are probably in the range of 4.5 to 5.5. The nature of watershed 111 was similar except there were several swampy areas.

The wells were drilled in glacial outwash (adjacent to location 6, Figure 3.1) which is currently intensively farmed and probably has been farmed for the last century. Considerable amounts of dairy manure are used along with some commercial fertilizer. Crops grown are generally corn for silage, legume hay, oats, and wheat. Probably the same practices have been followed for at least the last 25 years and perhaps the last century, although it is unlikely commercial fertilizer was used prior to 1925. The water level in the shallow wells was about 10 feet below the surface, which is a perched water table on a layer of till (sandwiched between porous outwash). These wells are adjacent to a field which received heavy applications of manure the preceding winter and which has almost certainly received considerable manure over the last several decades. The intermediate and deep wells are in the same outwash at different locations. They are adjacent to cultivated land which has been cultivated for several decades. Depth to water table was about 30 to 35 feet during the spring.

The averages of the observations are listed in Table 3.7 along with statistical parameters. The sampling interval for watershed 24 covered the period June 1973 to December 1974 and for watershed 111 from October to December 1974. The wells were sampled from April to December 1974. Discharge records for the streams are not available so that no good information on the influence of flow on concentrations is available. The concentration of MRP observed in watershed 24 tended to be somewhat higher in samples taken near what were maximum discharge rates, but this was not universally true. Probably the weighted mean concentrations would not be very different from the means listed in Table 3.7. The most remarkable feature of the data in Table 3.7 is the generally small variation in MRP and TSP among sampling sites and the relatively small standard deviation of the population. The well samples are not much different from the stream samples in either MRP or TSP, which

Table 3.7. Concentrations of TSP and MRP in water from streams draining subwatersheds with limited human activity and from wells in glacial outwash.

	MRP				*TSP*			
	Mean[1]	N[1]	S.D.[1]	S.E.[1]	Mean[1]	N[1]	S.D.[1]	S.E.[1]
Subwatershed 24	6.0	56	4.6	0.6	15	54	9.7	1.3
Subwatershed 111	5.2	9	2.9	1.0	24	6	11.9	4.7
Wells								
Shallow	5.4	28	2.3	0.4	12	22	5.9	1.2
Intermediate	6.7	9	3.5	1.2	20	8	5.3	1.9
Deep	7.1	21	3.4	0.7	13	17	4.1	1.0
All	6.2	58	3.1	0.4	14	47	5.9	8.6
Subwatershed + wells	6.0	123	3.9	0.3	15	107	8.6	0.9

[1] Mean expressed in μg P/l, N is number of observations.

S.D. is the standard deviation of the population, S.E. is the standard error of the mean.

suggests limited contamination of ground water from the manure and the stable nature of the SUP fraction.

The average of all observations summarized in Table 3.7 will be used as the estimates of concentrations of MRP and TSP which would be in the stream water if there was no human activity in the watershed. Some reservations about the validity of this are justified because human activity in watersheds 24 and 111 is absent because of geographical and geological factors (e.g., mantle of acid till, steep slopes, etc.); biogeochemical phosphorus may be influenced by some of these same properties, although there is no evidence that this is the case. The geological materials which most of the water contacted in watersheds 24 and 111 were acid while the water sampled in the shallow, intermediate and deep wells contacted materials considerably higher in pH, and probably the outwash materials at the depth of the intermediate and deep wells was calcareous. Thus water samples were derived from various geological materials and this seemed to make no important difference in phosphorus content.

Estimates of Phosphorus from Human Activities

Introduction. Human activity may generate phosphorus loading in various ways. Two such important ways are discharges of domestic sewage and milkhouse wastes. An example of the former is the effluent from a secondary sewage treatment plant at Dryden serving about 1400 people which is discharged directly into Virgil

Creek. Additional examples are overflows from lagoons serving trailer parks, improperly functioning home disposal systems and milkhouse wastes that are discharged directly into streams of all sizes throughout the watershed almost daily. Some domestic sewage and milkhouse wastes are discharged into the drainage system on a seasonal basis; during part of the year, the effluent from these systems is satisfactorily handled by the usual seepage field, but during excessively wet periods the seepage field may not function properly and the overflow is then discharged into drainage ways.

Figure 3.8. Roiled water in foreground is caused by inflow of effluent from the secondary sewage treatment plant in the background.

In addition to the above, there are a large number of sources which deliver phosphorus to streams only via overland flow of water. Their potential is not realized until excessive precipitation (that is, in excess of the ability of the soil to take in all of the precipitation) leads to overland flow. Examples are barnlots, manured fields, fertilized fields, etc. Within the watershed, the soils differ greatly in their ability to take in precipitation. Most soils in the watershed have a relatively high infiltration rate; that is, the surface of the soil is relatively permeable to water. However, some soils have relatively impermeable horizons or are shallow over bedrock and once the pores in the surface horizon are saturated they take in water at a relatively slow rate and overland flow occurs as a consequence. In other soils, which are very deep and permeable, the profile is seldom saturated and hence overland flow almost never occurs. Barnlots represent a special case in that the compaction of the animal traffic reduces the infiltration rate of the surface to very low values and hence the overland flow potential is very high.

Point sources. After a few initial studies, the influence of point source discharge of domestic sewage became evident, and considerable time and effort was spent in trying to determine how these point sources influenced the TSP loading at both low and high discharge rates. The Dryden sewage treatment plant is one such large point source and most of the effort has centered on it as an example. It is a secondary treatment plant which discharges effluent directly into Virgil Creek below Dryden (see Figure 3.1). Illustrated in Table 3.8 are results of one period of study in which detailed measurements of loading upstream and downstream were made during a low discharge rate period in the fall of 1973. The data in Table 3.8 conclusively demonstrate that the creek is a very poor conduit for soluble phosphorus at low discharge rate; the loading of TSP below the Dryden sewage treatment plant was 3.8 kg P/day while at Forest Home the loading from the whole watershed was only 0.7 kg P/day. There are many other point sources discharging phosphorus into Fall Creek in addition to the Dryden sewage treatment plant so the amount of soluble phosphorus added to the stream is considerably more than the 3.6 kg associated with the Dryden sewage treatment plant.

Table 3.8. Discharge, concentration and loading of phosphorus in Virgil and Fall Creeks for the period September 25–28, 1973.

Location[1]	Discharge[2] Rate, m³/sec	Concentration of MRP μg P/l	Loading, kg P/day	
			MRP	TSP
Above Dryden				
(4) + (21)	.20	5	0.1	0.2
Below Dryden (2)	.20	145	2.4	3.8
Virgil Creek at				
Freeville (16)	.24	83	1.7	2.3
Fall Creek at				
Forest Home (1)	.71	5	0.3	0.7

[1] Numbers in parentheses refer to locations in Figure 3.1.
[2] Discharge rate at 1 is greater than .71 m³/sec 85% of the time.

Probably the soluble phosphorus is removed from the stream primarily by reaction with the sediment in the bed of the stream. Biological reactions are an additional sink, but the vegetation in the creek itself is not very extensive and if the vegetation is to be an effective sink for the point source phosphorus from the Dryden sewage treatment plant as well as the other point sources, then considerable amounts would have to accumulate over the summer. Such accumulations appeared to be inadequate to account for more than a modest fraction of the observed removal of phosphorus.

Illustrated in Figure 3.9 are estimates of labile or sorbed P as determined by a sodium bicarbonate extraction procedure on sediments removed from the stream bed. These data show a large increase in labile P on the clay in the sediments just below the Dryden sewage treatment plant and a gradually decreasing labile P content with distance downstream, again substantiating the assertion that the sediments act as a sink for the point source P during periods of low discharge rate.

The results of these and other studies (e.g., Taylor and Kunishi, 1971) demonstrate that the sediments in the stream bed act as a sink for point source phosphorus during low discharge. The next question which arises is the influence of the phosphorus in the stream bed sediments during subsequent high discharge rates, when these sediments are swept up in the flow and transported either further downstream or maybe completely out of the watershed. If the reactions between the phosphorus and sediment are irreversible or only slowly reversible then the phosphorus stored in the stream bed sediment would have no appreciable influence on the soluble phosphorus of discharge during subsequent storms. However, if the reaction products behave as if the primary reaction were sorption, then the dissolution would be relatively rapid and the phosphorus stored in stream bed sediments would have an appreciable influence on the soluble phosphorus content of subsequent discharge during high discharge rate intervals.

Most of the evidence favors the view that a fraction of the point source phosphorus stored in stream bed sediments behaves as sorbed phosphorus and that it equilibrates within hours or days with solutions with which it is mixed. Several pieces of evidence support this assertion. First, the bicarbonate extraction data referred to above indicate large amounts of sorbed phosphorus on the stream bed sediments, as this procedure extracts an appreciable fraction of the sorbed phosphorus. The total amount of bed sediment in the stream was not determined.

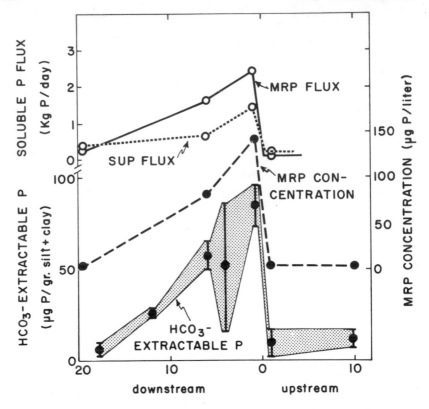

Figure 3.9. Bicarbonate extractable P per g of clay in bed sediments, MRP in
stream water and flux of MRP in stream flow plotted against distance upstream and downstream from Dryden for samples taken
during September and October of 1973. Points are means of 4 cores,
lines are drawn in for reference purposes, shaded portion surrounding bicarbonate extractable P line represents range of observations.

A second piece of evidence is the general decrease in MRP during
high discharge rate intervals at Forest Home for successive periods
of high discharge when these events occur frequently (fall-winter-
spring). This is illustrated in Table 3.6 by the mean MRP concentrations, which are generally highest in the first high discharge months
of the fall (November 1972, December 1973) and lowest in the late
spring months of March and April. The differences between the
1972–73 and 1973–74 high discharge seasons are also consistent
with the assertion that the phosphorus stored in the bed sediments

is effective in increasing the MRP concentration during high discharge intervals. In June 1973, phosphorus in laundry detergents
was banned in New York. This should have decreased the amount
of stored phosphorus in bed sediments derived from domestic sewage and hence reduced the effect of bed sediments on MRP in discharge during the fall and winter of 73–74 relative to the fall and
winter of 1972–73, which is consistent with the data in Table 3.6.

The above evidence indicates that some, but not necessarily all,
of the point source phosphorus which is removed from solution
during low discharge periods reappears as MRP during subsequent
high discharge. Unfortunately, this makes separation of phosphorus
derived from point sources and diffuse sources at high discharge
uncertain; some of the increase in concentration of MRP at high
discharge is a consequence of diffuse source inputs (via overland
flow) and some is a consequence of dissolution of sorbed P which is
derived from point sources during low discharge intervals.

Diffuse sources. One approach to separation of point source from
diffuse source phosphorus is described below. Using an empirical
method described by Chow (1964), hydrographs of selected high
discharge intervals at the Forest Home gaging station were separated into components which were attributed to surface flow and
subsurface flow (See Appendix B for details). The concentration
of MRP in the total flow was then plotted against the ratio of surface flow to total flow. By extrapolation of the trend of concentrations to 100% surface flow, an estimate of concentration in surface
flow was obtained. The results of this procedure are illustrated in
Figure 3.10. Evidently the winter of 1972–73 was different from the
winter of 1973–74. Two hypotheses may be advanced to explain
these differences. As pointed out previously, the phosphorus content of laundry detergent decreased in New York during the summer of 1973; this should reduce the amount of sorbed phosphorus
stored in the bed sediment prior to the high discharge events of the
winter of 1973–74 relative to those in 1972–73, and this difference
of sorbed P in the bed sediment could be a major factor influencing
the slope of the lines shown in Figure 3.10. This supposes that the
phosphorus from bed sediment and phosphorus derived from surface runoff are both important in determining the slopes of the
lines shown in Figure 3.10. The second hypothesis is based on the
supposition that vegetation undergoing decay in the soil surface is
an important source of phosphorus for surface runoff, particularly
early in the fall soon after senescence. The fall of 1972 was characterized by intense storms early in November which produced large

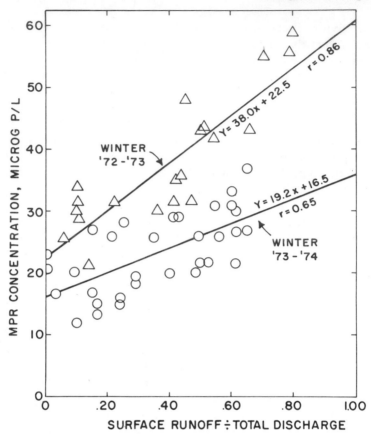

Figure 3.10. MRP concentration at Forest Home gaging station (location 1) plotted against the ratio of surface runoff to total discharge. The surface runoff component was determined from analysis of single peak hydrograph events using the method of Chow (1964); for the winter of 72–73, 5 storms were analyzed; for the winter of 73–74, 3 storms were analyzed. Points are experimental observations, lines represent regression equations.

amounts of surface runoff, while during the fall of 1973 no appreciable surface runoff occurred until late December. Hence the vegetation in 1972 would have been in the early stages of decay at the time of the first surface runoff event, while in 1973, the plant material would be in a more advanced stage of decay and much of the phosphorus would already have been leached into the soil by earlier rains not intense enough to produce surface runoff. The two effects discussed above confound each other and there is no evident way to separate them with the data we have available.

Another uncertainty is associated with the above procedure. The surface runoff derived from the hydrograph analysis is not necessarily all overland flow. The surface runoff by hydrograph analysis is that flow which reaches the stream quickly and at least a portion of this may in fact be derived from lateral flow within the soil. This water may emerge downslope in "seep" spots and then proceed to the stream via overland flow. Hence estimates of surface runoff by hydrograph analysis include the overland flow but are not necessarily composed entirely of overland flow.

Recognizing the limitations imposed by the above considerations, the effects of diffuse sources were estimated as follows. Extrapolation of the trends in Figure 3.10 were used to estimate concentrations of MRP in the surface runoff. Then the amount of total discharge which occurred as surface runoff was estimated for each year from an analysis of the hydrographs of high discharge events. Finally the effect of human activity on diffuse source loading was calculated as the product of the estimated concentration in surface runoff (corrected by the estimated concentration in the absence of human activity) and estimated surface runoff. The amount of surface runoff is estimated as 15% of the total discharge. The results are summarized in Table 3.9 for the experimental period. The contributions of diffuse sources for the periods not shown in Table 3.9 were considered zero since only small amounts of surface runoff occurred during these periods.

Table 3.9. Estimates of the effects of human activities on MRP loading from diffuse sources for the experimental period.

Period	Estimated Increase in Concentration of Surface Runoff, μg P/l	Estimated Surface Runoff m$^3 \times 10^6$	Loading kg P
November 1972–April 1973	55	24[1]	1300
December 1973–April 1974	28	17[1]	490
TOTAL			1790

[1] 15% of flow as surface flow.

The diffuse source loading from cropland was estimated by a second procedure and the results are summarized in Table 3.10. A description of this procedure follows. In all of the following, the approximations are generally such that the diffuse source loading is over-estimated. Approximately 10,000 ha of the watershed is devoted to cropland (hay, corn, small grains, etc.). Based on soil survey information and a sample of 20% of the farms, 40% of this cropland has a very low potential for overland flow, 34% has a medium potential for overland flow and 26% has a high potential for overland flow. Over the experimental period, total discharge averaged about 1 meter and of this total about 15 cm was surface flow according to the hydrograph analysis. Somewhat arbitrarily, this surface flow is distributed as 12 cm for the 34% of the cropland with medium potential and 30 cm for the 26% of the cropland with high potential for overland flow. Thus on 3400 ha of cropland, surface flow was about 1200 m³/ha and for 2600 ha of cropland surface flow was about 3000 m³/ha.

Table 3.10. Estimated diffuse source TSP from cropland.

Manure	Surface[1] Runoff Potential	Area, ha	Surface Runoff, m³/ha	Micrograms TSP/l	Loading, kg TSP
Yes	low	1200	0	—	—
Yes	medium	1050	1200	445	560
Yes	high	800	3000	445	1070
No	low	2800	0	—	—
No	medium	2350	1200	65	180
No	high	1700	3000	65	330
TOTAL					2140

[1] Low—Howard, Bath, Valois soil series.
 Medium—Mardin, Langford soil series.
 High—Lordstown, Volusia, Erie soil series.

Over the past six years extensive studies of TSP concentrations in surface flow from cropped plots treated with fertilizers and manure have been carried out at Aurora, N.Y. by P. J. Zwerman and his associates. These studies have been funded by EPA and a summary report is currently being prepared. Based on these data (personal communication, S. D. Klausner and P. J. Zwerman, Agronomy Department, Cornell University) the concentration of surface flow was estimated as 80 microgram/l TSP (standard deviation of population 20, standard error of mean 5) for unmanured (in the current year) cropland and 460 microgram/l TSP (standard deviation of population 103, standard error of mean 63) for cropland manured in the current year.

The manure from the 6800 dairy cows is assumed to be spread on 3100 ha of cropland and randomly distributed among the various drainage classes described above. Thus there are an estimated 1050 ha of manured cropland with medium potential for overland flow. Using the concentrations cited in the previous paragraph, the phosphorus loading from the cropland (in excess of biogeochemical phosphorus) is summarized in Table 3.10. Thus the cropland is estimated to have contributed about 2100 kg TSP during the experimental period, of which about 3/4 was derived from the manured cropland.

The estimates of diffuse source loading based on hydrograph analysis includes a fraction of the runoff from barnlots, and barnlot runoff is not included in the 2100 kg TSP estimated in the preceding paragraphs. Concentrations of phosphorus were measured in 35 samples taken within about 5 minutes of each other above and below a barnlot adjacent to sampling site 6. On the average, the samples below the barnlot contained 8 microgram/l more TSP than those taken from above the barnlot. (Standard deviation of population was 13, standard error of mean was 2). Since the watershed above the barnlot was 7 km² in area, the estimated total flow by the barnlot was 7×10^6 m³ of water during the experimental period and thus the loading was about 56 kg TSP (standard error ± 12 kg TSP) during the experimental period. This barnlot was adjacent to the stream, the only source of water for the animals was the stream itself, the feed bunkers were about 30 feet from the edge of the stream and the area of the lot was about 1 ha; overall the position and use of this barnlot was such as to nearly maximize loss of phosphorus to the stream.

In a study in Ohio, Edwards et al. (1972) measured loading from a beef feedlot. They found the average concentration of phosphorus in the surface runoff to be 6000 microgram/l of TSP during intervals when animals were in the lots. Total runoff amounted to about 30% of annual precipitation, but there were no animals in the feedlot from May to November. Using the concentration measured by Edwards et al. and estimating surface runoff at 1 M from the barnlot over the experimental period, yields an estimated loading of 60 kg TSP from the barnlot we measured, which is in agreement with its measured value.

Applying these results to the whole watershed, the average area of barnlot per cow is estimated as 40 m²/cow (430 ft²) for 80 cow herds (Chapter 5). For the 6800 cows in the watershed, this results in a loading of 1600 kg TSP (6000 microgram/l TSP, 1 M of runoff). Only about 15% of this is included in the diffuse source loading as calculated from the hydrograph analysis so about 250 kg should be added to the estimates of cropland loading to arrive at a value of about 2400 kg TSP for diffuse source loading.

The second procedure for estimating diffuse source loading agrees with the hydrograph analysis procedure; however, the two procedures are not entirely independent of each other and several arbitrary numbers are included in each calculation. However, both seem to be reasonable.

Correlation analysis was utilized as a third procedure for allocating the observed amounts of phosphorus in the stream among the original point and diffuse sources. During the period February to April 1973, measurements of MRP concentrations in several subwatersheds were made. The results are listed in Table 3.11 together with pertinent watershed characteristics. Listed in Tables 3.12 and 3.13 are correlations between selected variables. The correlation analysis yields no useful information about relative importance of several sources of phosphorus because of the high degree of correlation between area, cows, people, acres of cropland, etc. Several comments about the various subwatersheds seem warranted. First, the concentrations of MRP in 24, 61 and 9 were low primarily because of low densities of both people and cows. Second, the concentrations in 8 and 16 were high because of large point sources (relative to stream size); in 8, the milkhouse wastes and probably domestic sewage were draining into the stream, the barnlots were adjacent to streams and generally the farming operations utilized somewhat poorly drained soils. The results at 16 were high because of the Dryden sewage treatment plant. The remainder of the results fall in the range of 10 to 20 microgram/liter.

Summary of allocation of phosphorus among sources for the experimental period. Shown in Table 3.14 are estimates of the contributions of the various sources to MRP, TSP and labile P loading during the experimental period. The amounts of MRP and TSP expected if all human activity were removed from the watershed were calculated as the product of the appropriate concentrations derived in the section on biogeochemical phosphorus and total flow for the experimental period. The point source inputs attributed to human activity were estimated by difference. The labile P was apportioned among the various sources in the same ratio as the MRP.

The results show that approximately 50% of the TSP loading is associated with human activity. Diffuse and point sources associated with human activity each contribute about 1/4 of the TSP.

The uncertainties associated with the numbers in Table 3.14 vary considerably. The total MRP and TSP loading is reasonably well

Table 3.11. Average concentrations of MRP during the period February through April 1973 for subwatersheds. Standard errors of means varied but were usually less than 5 μg/l.

Sample Site No.	MRP μg P/l	Area, ha	Cows, Number	Active Agr, ha	Corn, ha	Houses, Number
24	6	229	0	13	0	0
61	8	664	8	124	4	21
21	16	2676	350	930	181	192
9	10	554	26	91	4	16
12	14	1282	326	608	138	135
11	15	940	162	350	65	27
8	50	272	126	137	29	9
18	18	362	0	56	12	13
10	18	3130	608	1261	253	215
7	16	8081	742	1981	293	251
4	12	4931	1175	2152	428	317
3	20	1403	387	628	172	170
13	16	2221	543	1042	190	—
15	15	14745	4332	6140	1800	—
16	30	10700	2099	4436	953	—
1	19	33570	8852	12435	2985	—

Table 3.12. Correlation of MRP and (MRP) (area) with selected variables.

	Group	MRP			MRP $\times$ area		
		1[1]	2[1]	1+2	1[1]	2[1]	1+2
Area		—.12	—.56	—.20	.97	.92	.96
Cows		.06	—.58	—.10	.96	.76	.91
Corn		—.02	—.55	—.12	.94	.82	.94
Houses		—.11	—.51	—.17	.90	.87	.95

[1] Group 1 locations are generally the smaller watersheds and include sampling locations 24, 6, 9, 12, 11, 8, 18, 10, 13, 3; Group 2 are generally the larger watersheds and include sampling locations 21, 7, 10, 4, 13, 3, 12, Table 3.11.

documented and the standard error of the estimates is on the order of 5% of the estimates. The estimates of biogeochemical phosphorus are subject to considerable uncertainty simply because human activities preclude making estimates from a wider range of sites. The variability among the sites which were measured is remarkably low. The influence of human activity on diffuse source loading is again very uncertain because of the uncertainty in separating point sources and diffuse sources using the hydrograph analysis. However, independent estimates of diffuse source inputs furnish reasonable confirmation of the results of the hydrograph

Table 3.13. Correlation matrix of characteristics of the subwatersheds listed in Table 3.11.

	Total Area	Cows	Area of Alfalfa	Area of Corn	Area of Inactive Agriculture	Houses
MRP	−.20	−.10	−.15	−.26	−.12	−.17
(MRP) (Area)	.96	.91	.89	.95	.94	.95
Total area	1.00	.96	.95	.97	.99	.96
Cows	.96	1.00	.99	.99	.92	.95
Area of alfalfa	.95	.99	1.00	.98	.91	.93
Area of corn	.96	.99	.98	1.00	.94	.98
Area of inactive agriculture	.99	.92	.91	.94	1.00	.94
Houses	.96	.95	.93	.98	.94	1.00

analysis. Finally, the point source estimates are the most uncertain because they are calculated as the remainder; that is, total minus biogeochemical and diffuse sources.

One important limitation imposed on any "tinkering" which might be done with the estimates attributed to the various sources is that the sum of biogeochemical, diffuse and point sources must equal the measured loading; an increase from one source must be balanced by a decrease in another, etc.

Listed in Table 3.15 are estimates of phosphorus which is cycling through the farming operations each year and which has potential for serving as a diffuse source of TSP and labile phosphorus. The total amount of such sources is about 450,000 kg P for the experimental period; approximately 0.5% of this phosphorus was transported out of the watershed as TSP. In general the phosphorus from the diffuse sources is retained reasonably well within the watershed.

Domestic sewage from the roughly 12,000 inhabitants in the watershed probably contained about 1.5 kg P/cap/yr for the period prior to about June 1, 1973 and about 0.9 kg P/cap/yr after June 1, 1973. Total phosphorus in domestic sewage during the experimental period amounted to about 25,000 kg P of which about 3200 kg or 13% was estimated to leave the watershed as TSP.

The average export values are not representative of all situations. Manure spread on certain soils may have contributed nothing while perhaps 10% of the soluble phosphorus from other manure may have been lost to surface runoff. With respect to domestic sewage, the same situation applies; many single home disposal systems

Table 3.14. Estimated contributions of various sources of MRP, TSP and labile P for the experimental period September 1972 through April 1974.

Source	MRP		TSP		Labile		TSP + Labile	
	kg P	%	kg P	%	kg P	%	kg P	%
Biogeochemical	2000	33	5000	50	400	33	5,300	48
Human	4000	67	5000	50	800	67	5,800	52
Diffuse	1800	30	2200	22	400	30	2600	23
Point	2200	37	2800	28	400	37	3200	29
TOTAL	6000	100	10000	100	1200	100	11,100	

Table 3.15. Potential diffuse sources of TSP and labile P in the Fall Creek watershed for the period September 1972 through April 1974.

Source	Method of Estimation	Total, kg P
Dairy manure	6800 cows[1], 35 kg P/cow	240,000
Fertilizer	Survey[1]	180,000
Precipitation	Measurements at Aurora, N.Y., 1973[2]	21,000
TOTAL		441,000

$$\frac{\text{Diffuse source loading of TSP} + \text{Labile P}}{\text{Potential sources}} = \frac{2600}{441,000} \times 100 = 0.5\%$$

[1] J. J. Jacobs, unpublished data.
[2] S. D. Klausner, unpublished data.

work well and essentially all of the phosphorus is removed from the effluent as it percolates through the soil, but the remainder work imperfectly and some considerable fraction of the effluent is discharged to a stream before the soil has removed phosphorus. Many multiple residence systems such as the Dryden sewage treatment plant or lagoon systems for trailer parks work satisfactorily so far as BOD and pathogen removal are concerned, but the phosphorus is not removed, and it is discharged into streams.

Estimates based on measurements made in Virgil Creek adjacent to Dryden indicate inputs from Dryden total about 2800 kg during the experimental period. Based on the estimate of 3200 kg in Table 3.14 other point sources contributed only about 400 kg P, which is unreasonable since we know there are a large number of other point sources in the watershed. In addition, according to our analysis of diffuse source inputs, about 1400 kg P from barnlots was included in the point source category because of the nature of the hydrograph analysis. Thus, the amount of point source TSP appears to be larger than the estimate derived by difference. The most likely explanation for this discrepancy is transmission loss—that is, losses which occur after the phosphorus is placed in the stream. Other information reviewed in the succeeding paragraphs indicates transport losses may be considerable.

Illustrated in Figure 3.11 are estimated weighted mean concentrations and loadings calculated for various points in the watershed for the period May 1, 1973 to April 30, 1974. Numbers in parentheses are estimated on the basis of average values for the watersheds above locations 4, 21 and 15. While there are a number of unmeasured inputs into the network and the standard errors of the estimated concentrations are in the range of 10 to 20%, overall, the

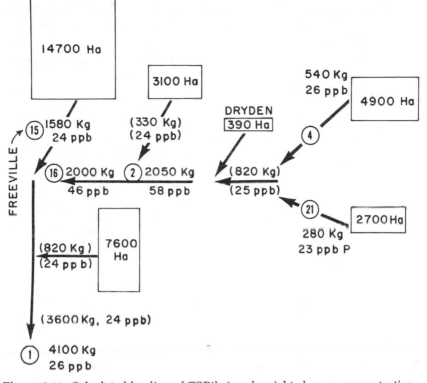

Figure 3.11. Calculated loading of TSP(kg) and weighted mean concentration, ppb (which is equivalent to μg P/l), for subwatersheds in Fall Creek basin, May 1, 1973 through April 30, 1974. Numbers without parentheses are measured values, numbers in parentheses are estimated as sum of converging streams or are estimated from weighted mean concentrations of locations 4, 21 and 15. Numbers in circles are sampling locations, Figure 3.1.

impression is one of appreciable transmission losses. For subwatersheds above 4, 21 and 15, the weighted mean concentration averaged 24 μg P/l for the period May 1973 to April 1974 (these subwatersheds represent over 2/3 of the area of the whole watershed); assuming this would be the concentrations observed in the remaining 1/3 of the watershed were the Dryden sewage treatment plant not operating, a loading of 3600 kg TSP is calculated. Since 4100 kg was observed, the estimated effect of Dryden was 500 kg P, compared to estimated inputs of 1200 kg P for the same period, or transmission losses of roughly 60%. Looked at another way, if the concentrations at locations 15 and 16 are weighted by area and the output from the watershed below Freeville is set equal to 24 μg

P/l, the calculated weighted mean concentration at Forest Home would be 31 μg P/l, while the measured concentration was 26 μg P/l. Thus the "transmission" loss of TSP from the Dryden sewage treatment plant appears to be on the order of 50% or more. An additional piece of evidence is found in Table 3.12, where generally MRP concentrations decreased as watershed size increased as indicated by the negative correlation of MRP on watershed size even though human activities per unit area remained constant.

Applying a 50% transmission loss generally to the estimated point source inputs, point source inputs (before transmission losses) of about 2400 kg P for the period May 1973 to April 1974 are calculated. During this period, estimated losses from barnlots not included in diffuse sources were 700 kg and inputs from the Dryden sewage treatment plant were 1200. Thus all other domestic sewage point sources amount to about 500 kg P. For roughly 10,000 people this is about 0.05 kg P/cap/year, of which only 0.025 kg P/cap/year is delivered out of the watershed. Thus on the order of 3% of the phosphorus of unsewered populations is delivered out of the watershed as TSP. Clearly, location of the point sources is likely to be very important; the TSP from point sources close to the stream outlet would undergo much less transmission losses than those more distant (location 2 vs. location 16 vs. location 1, for example). There are numerous point sources in watersheds 4, 21 and 15 (e.g., location 8) and they have some influence on the observed concentrations, but the "weak" point sources are diluted out and the phosphorus is removed by (we hypothesize) reactions with the sediments and by incorporation into organic matter during transmission. Probably transmission losses are highly important factors.

Extrapolation to other time periods. So far attention has been directed to analysis of the results for the experimental period only; in this section the results are extrapolated to other time periods. First, the differences between years ('72–'73 vs. '73–'74) present serious problems because these differences have been shown to be consistent with two hypotheses: (a) the ban in June 1973 on phosphorus in laundry detergents and (b) the severe storms of the fall of 1972. Acceptance of the former hypothesis would imply that the '73–'74 results are indicative of the present and the '72–'73 results are indicative of the past. Acceptance of the latter hypothesis would imply that the two years be combined with appropriate weight given the years according to probability of occurrence based on past flow records. The following discussion indicates both hypotheses are of some consequence, but probably the detergent ban is of most importance so far as extrapolation through time is concerned.

The sources of TSP for the periods September 1972 to September 1973 and May 1973 to April 1974 are examined in Table 3.16. These results indicate the data are reasonably consistent with the effects expected from the detergent ban and 1800 kg of the 1972–73 loading can be attributed to domestic sewage while the corresponding figure for 1973–74 is 1200 kg.

Table 3.16. Contributions of various sources to TSP loading for the period September 1972 through April 1974 at location 1.

September 1972 through August 1973	
(1) Estimated TSP (Table 3.6)	6540 kg
(2) Total discharge	206.3 $\times$ 10^6 M^3
(3) Biogeochemical TSP (discharge $\times$ 15 μg P/l	3100 kg
(4) Human activity (1) — (3)	3440 kg
(5) Diffuse source TSP (Table 3.9)[1]	1600 kg
(6) Point source TSP, (4) — (5)	1840 kg
May 1973 through April 1974	
(7) Estimated TSP (Table 3.6)	4110 kg
(8) Flow	155 $\times$ 10^6 M^3
(9) Biogeochemical TSP (discharge $\times$ 15 μg P/l	2320 kg
(10) Human activity (7) — (9)	1790 kg
(11) Diffuse source P (Table 3.9)[1]	600 kg
(12) Point source P, (10) — (11)	1190 kg
(12)/(6) = 0.65; estimated ratio of sewage P/cap[2]/yr = 0.70	

[1] Calculated from ratio $\dfrac{\text{MRP, diffuse source}}{\text{MRP, human activity}} = \dfrac{\text{TSP, diffuse source}}{\text{TSP, human activity}}$

[2] 1.5 kg P/cap/yr for September through July 1973, 0.9 kg P/cap/year thereafter.

Examination of the flow records from 1939 to 1971 show that in the 33 Novembers in this period, in only one case did daily flow exceed 42 m^3/sec (1500 cfs) and on only 9 days (7 separate months) did flow exceed 28 m^3/sec (1000 cfs); in 1972, daily flow exceeded 42 m^3/sec one day and 28 m^3/sec 2 days. Hence only about 7 out of 33 Novembers would we expect 28 m^3/sec and only 1 out of 33 Novembers would we expect 42 m^3/sec. Thus if the explanation for the difference between '72–'73 and '73–'74 rests on the nature of the November storms, then not much weight should be given the November 1972 results because such Novembers occur infrequently.

The preceding discussion suggests that the 1973–74 data is the best indicator of the future regardless of what weight is given to the two hypotheses listed above. Furthermore, the portion of the difference attributed to the detergent ban in Table 3.16 should be used as an indicator of differences between the past and future.

Several schemes for extrapolation of results to other years in Fall Creek can be visualized. The simplest procedure is to multiply the weighted mean concentration by the total discharge measured or expected during the year. For example, total discharge averaged over the last 33 years was 50 cm per year and weighted mean concentration for 1973–74 was 26 μg P/l (when total discharge was 46 cm). Thus average loading in the future would be on the order of 4400 kg TSP, if the weighted mean concentration of TSP during 1973–74 is a reasonable estimate of present and future conditions. A second procedure could be used which would take into account the characteristics of the various sources. For example, point source inputs should remain relatively constant from year to year (barring major changes in phosphorus content of sewage) while diffuse source inputs and biogeochemical inputs should increase as flow increases. Using this second procedure, past loading in Fall Creek (before detergent ban) is estimated as different from the future loading by the difference in point source loading of 600 kg P/year shown in Table 3.16. Thus average loading in Fall Creek prior to June 1973 is estimated as 5000 kg TSP/year.

Extrapolation to other watersheds. In this section the results of the previous research are extrapolated to other watersheds. Several limits and conceptual ideas were imposed on the generalizations presented here. These are as follows:

(1) The results are summarized with respect to the effects the phosphorus is likely to have on the lakes in central New York discussed in the previous chapter. There, the correlations between lake characteristics and watershed variables were based on a literature review and some of the preliminary results from the Fall Creek studies. Here, the same lake characteristics will be used as a means of evaluating the extrapolations based on the Fall Creek studies and literature values.

(2) In the previous chapter, the effects of drainage basin characteristics were summarized in terms of units of phosphorus per unit of basin characteristic (for example, 1.5 kg of P per capita of sewered population). This same procedure is followed here to some extent except that the only four basin characteristics considered are sewered population, unsewered population, total area and area of active agriculture. The remainder of the basin characteristics were shown in the previous chapter to be relatively unimportant in terms of soluble P loading.

(3) The variability of the inputs of the sources among basins is not well documented and a brief discussion of the uncertainties follows.

The per capita phosphorus content of domestic sewage should be relatively constant over the watersheds considered and the removal of phosphorus from the collected sewage is probably fairly constant. The discharge and delivery of TSP from the unsewered populations is an uncertain parameter. The range of individual systems is likely from zero to 100% delivery. Average regional differences may be fairly large; in those areas where the unconsolidated mantle is permeable and well drained, delivery is nearly zero while in those regions with poorly drained soils the delivery may range up to 100% where villages are adjacent to lakes or on streams only a short distance from the lakes.

The watersheds within the delineated area are reasonably similar. Dairy farming is the major enterprise in most. The most intensively farmed soils have many important characteristics in common.

While the ideal of a constant loading per unit of watershed characteristic is not achieved, we will proceed with the working hypothesis that the loading per unit basin characteristic can be approximated by a constant. This working hypothesis will be examined in relation to its consistency with respect to lake parameters and how sensitive this type of test is in regard to furnishing unbiased estimates of reliability.

(4) The concentration of TSP in stream flow is postulated to be relatively constant for all forest land across the region, a different concentration is characteristic of farmed land, etc. Thus loading per unit of land area is not a constant, but is proportional to total stream discharge. In the central New York region under consideration the average total discharge per unit of land area ranges from about 0.3 m/yr in the western section to 0.6 m/yr in the eastern section. Thus the loading per unit of land area is postulated to double from west to east across the region considered because of the difference in average annual discharge. This postulate is consistent with the data from the central New York lakes as illustrated in Table 3.17 by the correlations between winter phosphorus in lakes, area of land use per unit of lake surface area, and product of area of land use per unit of lake area times discharge per unit of area.

(5) Input per capita of sewered and unsewered populations were approximated by per capita per year constants regardless of location with respect to the lake.

Table 3.17. Correlation between winter TP in lakes in central N.Y., land use per unit of lake area, and product of land use per unit of lake area and discharge per unit of land area.

	Correlation Coefficient of Winter TP with	
Land Use Category	Land Use	Land Use × Discharge
Active agriculture	.51	.67
Forest plus		
inactive agriculture	.40	.76
Residential area	.89	.92

(6) Based on the above considerations the model for loading per unit of lake surface area becomes

$$L_{fc} = a_1DT + a_2DA + a_3P_s + a_4P_u \qquad (3.4)$$

where,

L_{fc} is loading of P (kg P/ha lake surface/yr) based on Fall Creek parameters

a_1 is concentration of biogeochemical TSP

D is total discharge per unit of land surface per year

T is total land area per unit of lake surface area

a_2 is concentration of TSP from agricultural land in excess of biogeochemical TSP

A is area of active agriculture per unit of lake surface area

a_3 is input of phosphorus per capita of sewered population per year

P_s is number of sewered population per unit of lake surface area

a_4 is input of phosphorus per capita of unsewered population per year

P_u is number of unsewered population per unit of lake surface area

(7) The constants a_1 through a_4 were assigned values as follows. First, somewhat arbitrarily, the input of the sewered populations was set equal to 1.0 kg P/cap/yr. The phosphorus content of domestic sewage was probably in the range of 1.5 kg per cap per year prior to the ban on phosphorus in laundry detergents in New York; the lower figure reflects removal in sewage treatment plants and transmission losses between the sewage treatment plant and lakes. Second, the input of unsewered populations was set equal to one-tenth this or 0.1 kg P/cap/yr. This is somewhat higher than that estimated for Fall Creek, but probably the Fall Creek estimates are

low because of large transmission losses. Third, the biogeochemical TSP was estimated as 15 μg P/l. Fourth, the coefficient for farmed land was estimated as follows. As summarized in Table 3.9 TSP from diffuse sources during the period September 1972 through April 1974 was 2200 kg TSP (in excess of biogeochemical phosphorus). This was derived from 124 km^2 of active agricultural land and it was contained in 124×10^6 M^3 of H$_2$O. Hence average concentration in excess of biogeochemical TSP was 18 μg TSP/l.

Based on the above limits and approximation, values of L_{fc} were calculated for each of the 13 lakes discussed in the previous chapter. The correlation between L_{fc} and TP in the lakes was 0.83 and between chlorophyll in the lakes and L_{fc} was 0.70, which are the same as those reported in the previous chapter for loading based on somewhat different considerations.

The correlations listed above indicate the coefficients derived from the Fall Creek studies are consistent with the differences among the lakes in central New York. However, a more careful examination of the data indicates that perhaps the correlations between lake parameters and calculated loading are somewhat misleading. Listed in Table 3.18 is the correlation matrix for the vari-

Table 3.18. Correlation matrix for selected lake and lake basin characteristics (data from Chapter 2).

		X$_1$	X$_2$	X$_3$	X$_4$
TP in lakes	(TP)[1]	.61	.80	.71	.55
Active agriculture	X$_1$[2]	1.00	.64	.63	.07
Total area	X$_2$[3]		1.00	.36	.49
Sewered population	X$_3$[4]			1.00	.46
Unsewered population	X$_4$[5]				1.00

[1] Total phosphorus in lakes in winter (see Chapter 2).

[2] Product of active agriculture (ha per ha of lake surface) and average stream discharge per unit of land area.

[3] Product of total area (ha per ha of lake surface) and average stream discharge per unit of land area.

[4] Sewered population per unit of lake surface area.

[5] Unsewered population per unit of lake surface area.

ables under consideration. As illustrated in this table, the so-called independent variables are fairly well correlated with each other and hence this correlation suggests a wide range of values could be assigned to a_1 through a_4 and the values of L_{fc} would be well correlated with the lake parameters. This was illustrated in the following manner. Fifteen sets of values of a_1 through a_4 were derived from a random number table (0 to 100) and these 15 sets of random numbers were used to calculate 15 sets of L_{fc}. Each set of L_{fc} was then correlated with the measured TP values. Eight of the 15 correlation coefficients were in the range of 0.74 to 0.76 (as compared to 0.83 for L_{fc}) and the remainder ranged downward to 0.58. Thus the coefficients derived from the Fall Creek results or derived in the previous chapter from the literature were slightly better than one-half of the loading coefficients calculated from random numbers. This illustrates that the high degree of correlation among variables biases the statistical statements about the relative importance of the variables.

A second method of analysis follows. The multiple regression of TP on sewered populations, unsewered population, total area and active agricultural areas (all expressed per unit of lake surface area) was calculated and is as follows:

$$\text{TP} = 1.92 + 0.0586\ P_s + 0.0094\ P_u + 3.51\ DT - 4.11\ DA$$
$$\qquad\qquad (\pm0.023)\quad (\pm0.030)\quad (\pm1.18)\quad (\pm2.57)$$

$$R = 0.9 \tag{3.5}$$

where,

   TP = average concentration of phosphorus in lakes in winter, mg/m^3

   P_s = sewered population per km^2 of lake surface

   P_u = unsewered population per km^2 of lake surface

   D = discharge per year per unit of land area

   T = total land area, ha per ha of lake surface

   A = Active agriculture, ha per ha of lake surface

   Numbers in parentheses below coefficients are standard errors of coefficients.

A disconcerting aspect of the above equation is the negative coefficient for active agriculture. A number of statistical studies were carried out with these and other variables (residential area as a proxy for population, discharge per unit of lake surface area, etc.). In all cases where some measure of population was included, the coefficient for active agriculture was negative. Using residential area, active agriculture and forest + inactive agriculture as variables, 12 regressions were fitted leaving out the data from one lake in order. In all cases, the coefficients were much the same and the coefficients for agriculture were negative.

Several hypotheses can be advanced to explain the negative coefficient for agriculture. Clearly, all that can be done with the data at hand is *suggest hypotheses for further testing* and the following are some of the hypotheses that have occurred to us:

(a) The negative coefficient is a consequence of the correlation among the independent variables.

(b) As active agriculture increases, the fraction of the area with well-drained soils increases; as a consequence, sewage disposal fields for home systems work better and hence there is less input from domestic sewage.

(c) As active agriculture increases, the population is more dispersed and on the average further from the lake; a larger fraction of the domestic sewage which is placed in the drainage network is converted to biologically inactive form during transmission.

(d) As active agriculture increases, cash crop production may increase and hence the mix of manured and unmanured land would change.

Thus although the negative coefficient for active agriculture is perhaps a consequence of intercorrelation among so-called independent variables, it may also be a consequence of real effects which should be investigated more carefully.

All of the above discussion leads to one inescapable conclusion; namely, the estimates of phosphorus inputs associated with human activities are uncertain and we cannot assign formal limits to the uncertainty on the basis of conventional statistical analysis. A distinction should be made between the value of correlations as a guide to interpolation and as a guide to management. The correlations and regressions are much more reliable as interpolation guides than as management guides; for interpolation purposes the correlations and regressions would be used to predict the state of lakes in central New York, given the basin characteristics. In the case where inferences about management are to be made, statements are desired about how removal of phosphorus from domestic sewage or control of barnlot runoff would influence lake quality. This latter interpretation is much more uncertain than the former. Correlations do not prove cause and effect relationships, while management decisions are based on what are perceived to be cause and effect relationships. Surely management decisions should not be based solely on correlation analysis when the so called independent variables are as highly correlated as those discussed here. Furthermore, probably the correlations reported here are typical of most basins.

Despite all of this uncertainty, several conclusions seem firmly established both experimentally and statistically. First, the input of phosphorus from domestic sewage from sewage treatment plants whose output is fed into the lake or into streams a short distance from the lake are well documented by direct evidence from the literature. For inputs a considerable distance upstream from the lake, the transmission losses introduce an element of uncertainty; losses likely vary considerably among stream reaches and among streams.

The input from unsewered populations is subject to much more uncertainty. First, if the disposal field is properly designed and works well then probably the input is essentially zero; most of the phosphorus is removed in moving through the soil. Second if the disposal field does not perform properly then the overflow usually drains into a road ditch or some other drainageway and transmission losses can be appreciable. The disposal field probably functions properly at least a part of the year so that not all of the phosphorus is discharged even with poorly functioning systems. The importance of the unsewered population must vary greatly among watersheds; in those watersheds with relatively well drained soils, probably inputs of phosphorus are no more than 1% of the total phosphorus in the sewage from the unsewered population. In watersheds with more poorly drained soils, the corresponding inputs may range upwards from 10%, although probably if inputs are in this range the health hazards, odors, and aesthetic considerations will lead to sewer construction. Furthermore, transmission losses further attenuate the influence of the unsewered populations which are at a considerable distance from the lake. On the other hand, lakeside cottages with poorly functioning disposal fields are special hazards because the transmission losses approach zero. Probably on the order of 0 to 50% of the phosphorus from unsewered populations is transported to lakes if a wide range of conditions is surveyed.

The concentrations of TSP in runoff from forested land appears to be about 15 μg P/l. This is the concentration estimated from the Fall Creek studies, and it is the concentration of phosphorus measured by Taylor et al. (1971) for a forested watershed in Ohio using procedures similar to those we have used. (Relatively few studies can be compared with those we have carried out because many investigators have included some or all of the particulate phosphorus in their loading calculations.)

The influence of farming operations on the composition of stream water is likely to be somewhat variable. In the preceding discussion, the effect of farmland was estimated as the product of concentration times overland flow. Concentrations of phosphorus in barnlot runoff were estimated as 6000 μg TSP/l. Based on plot studies at Aurora, New York (Zwerman *et al.*, personal communication) the concentration of phosphorus in surface runoff from plots unmanured in recent years, from plots manured the previous year, and from plots manured in the current year was about 40, 80 and 450 μg TSP/l, respectively. The composition of the runoff was extremely variable and the standard deviation of the populations is on the order of one hundred percent. The amount of overland flow is a complex function of location of the soil in the landscape, the nature of the soil profile, the nature of the precipitation event and soil management factors. Farming tends to be concentrated on the better drained soils and these soils are also least subject to overland flow. The concentrations referred to above illustrate the overriding importance of manure on concentrations in overland flow. Thus the proportion of dairy farms relative to other enterprises may be very important, particularly since dairy farmers can utilize some of the more poorly drained soils fairly effectively.

The fraction of the total discharge from a watershed that reaches streams as surface flow (based on hydrograph analysis) is likely to be on the order of 5 to 20% of the total discharge in the region of New York under consideration; the largest fraction of this is likely to be derived from the more poorly drained soils and most of these will usually be in inactive agriculture and brush and forest land. Thus the crucial factors for farming are (a) barnlots and (b) the amount of manure applied to the more poorly drained soils.

Transmission losses for phosphorus in runoff from barnlots and agricultural land are probably fairly low; the overland flow generally occurs only when flow is high so that transport times are short, the more reactive suspended solids are moving with the water and generally the time for either chemical precipitation or biological incorporation is too short to have a major impact.

Summary of Phosphorus Studies

1. In Fall Creek, the concentration of soluble phosphorus was about 30 micrograms per liter and the particulate matter contained about 110 μg P/l. Most of the latter phosphorus was carried out of the watershed during the short intervals of time when the discharge rate was very high.
2. The soluble phosphorus is probably the form most important to the biology of the lakes.
3. About 50% of the soluble phosphorus was attributed to non-

human activities, about 25% was attributed to domestic sewage, and about 25% was attributed to farming operations.

4. Based on the Fall Creek data and numerous other considerations, the loading of soluble phosphorus to lakes in central New York can be approximated by the sum of the following inputs:
   a. Sewered populations:
      (i) phosphorus in laundry detergents 1.0($\pm$0.5) kg P/cap/yr
      (ii) No phosphorus in laundry detergent 0.5($\pm$0.4) kg P/cap/yr
   b. Unsewered population: 0.1 to 0.4 of the values for sewered population on a per cap/yr basis
   c. Nonagricultural land: 15D mg/m^2/yr, where D is meters of stream flow per m^2/year. Probably the range is 10D to 20D.
   d. Agricultural land (in excess of 15R): 18D mg/m^2/yr, where D is meters of stream flow per m^2/yr. Probably the range is 10D to some unknown upper limit.

NITROGEN STUDIES IN FALL CREEK

Introduction

The nitrogen studies in Fall Creek concerned NO_3-N. In the lakes in central New York algal growth does not appear to be limited in any important way by nitrogen, so the influence of nitrogen availability to aquatic organisms seems to be unimportant here. The subject of major interest so far as nitrogen is concerned is reduced to possible toxic levels of NO_3-N and sources of nitrate N. The NH_4-N content of all samples was less than 0.3 mg. N/l.

Procedures

The samples described in the previous sections were analyzed for NO_3-N. The supernatant from the suspended solids determination was analyzed for nitrate by a steam distillation procedure (Bremner, 1965) in which NO_3-N was reduced to NH_4-N by Devarda's alloy in the presence of $Mg(OH)_2$ and the distilled $NH_4{}^+$-N titrated with standard acid (NH_4-N was determined prior to NO_3 analysis by the same steam distillation procedure).

Results

About 2500 samples from several subwatersheds were analyzed for NO_3-N over a period of 24 months. Concentrations were less than 5 mg NO₃-N/l in all samples; in one sample the concentration exceeded 4 mg NO₃-N/l and in five samples the concentration exceeded 3 mg N/l. Thus based on the PHS recommended maximum concentration of 10 mg NO₃-N/l for drinking water, no samples were found any place at any time in which the NO₃-N concentration was a health hazard.

Generally the concentrations of nitrate N did not change much with differing discharge rate during the course of a week or so, and regressions of NO₃-N on discharge rate usually explained only a small fraction of the variation. Hence the data were summarized as monthly averages.

Average nitrate nitrogen concentrations in samples taken from Fall Creek near the Forest Home gaging station are listed in Table 3.19 by months. Also listed are averages reported by Likens (1974) for samples taken in 1970 and 1971 at a sampling station about 4 km below the gaging station. For a given month, the means are remarkably similar among years. Within a given month, the individual values were not much influenced by discharge rate as indicated by the standard errors of the means. Likewise the total monthly discharge did not appear to influence average concentrations; for example, total discharge in November 1972 was 28×10^6 M³ with an average concentration of 0.70 mg NO₃-N/l and in November 1973 total discharge was 3.2×10^6 M³ with an average concentration of 0.72 mg NO₃-N/l.

The seasonal pattern of nitrate nitrogen concentrations was similar among the years with minimum concentrations occurring during the summer and maximum concentrations occurring during the winter. In order to summarize the data for a given sampling location with one equation, a regression of the form,

$$Y = a + b \sin 2\pi\left(\frac{m}{12}\right) \qquad (3.6)$$

where

$Y =$ monthly mean concentration of NO₃-N, mg/l
$m =$ index of month, with
$m = 0$ for Nov., $m = 1$ for Dec., etc.
a and b are regression coefficients

was fitted to the data. Using the monthly means listed in Table 3.19 the R^2 value for the regression was 0.89. The predicted values and their standard errors are listed in the last column of Table 3.19.

Since the mean monthly concentrations were not strongly influ-

Table 3.19. Average nitrate nitrogen concentration in Fall Creek samples taken near the Forest Home gaging station.

Month	1972 Mean	N[1]	S.E.[2]	1973 Mean	N	S.E.	1974 Mean	N	S.E.	1970[3] Mean	1971[3] Mean	Predicted[4]
January				1.63	10	.09	1.38	42	.04		1.65	1.44 ± 0.18
February				1.36	7	.09	1.59	39	.07		1.47	1.53 ± 0.18
March				1.44	3	.06	1.44	40	.02		1.75	1.44 ± 0.18
April				0.93	8	.07	1.27	84	.02		1.47	1.22 ± 0.18
May				0.73	15	.05	0.64	28	.02		0.84	0.90 ± 0.18
June				0.82	12	.05					0.66	0.58 ± 0.17
July				0.52	8	.06					0.42	0.36 ± 0.17
August				0.28	6	.04				0.30		0.27 ± 0.17
September				0.41	46	.03				0.34		0.36 ± 0.17
October				0.19	51	.02				0.64		0.58 ± 0.17
November	0.70	13	.07	0.72	48	.02				0.90		0.90 ± 0.18
December	1.16	19	.09	1.17	78	.04				1.28		1.22 ± 0.18

[1] Number of observations.

[2] Standard error of mean.

[3] Likens (1974).

[4] Based on regression $NO_3\text{-}N = 0.90 + 0.63 \left[\mathrm{Sin}\, 2\pi \left(\dfrac{m}{12} \right) \right]$

where m = 0 for November, 1 for December, 2 for January . . . 11 for October.

enced by total discharge, loading was calculated as the product of monthly discharge and mean monthly concentration. Furthermore, an average loading was estimated using the average monthly discharge based on the 1943–1972 averages. The results of these calculations are shown in Table 3.20 together with estimated standard errors. The yearly total of 5.5 kg NO_3-N/ha represents the average loading expected in Fall Creek over a long period of time, assuming inputs of nitrogen do not change drastically.

The nitrate data from several subwatersheds were extensive enough that the seasonal regression model could be fit to the data. Listed in Table 3.21 are the regression constants for several subwatersheds together with loading calculated on the basis of the average monthly runoff per unit area at the Forest Home gaging station. Watershed 24 serves as a measure of the amount of NO_3-N which is lost in the absence of human activity. The loss of 1.0 kg N/ha/yr is comparable to the value of 1.25 kg N/ha/yr reported by Bormann et al. (1968) for an undisturbed, forested watershed at Hubbard Brook, N.H.

Table 3.20. Nitrate nitrogen concentration and loading for Fall Creek at the Forest Home gaging station, based on concentrations calculated from measurements made during the period 1970–1974 and average flow, based on 1943–1973 data.

Month	Flow[1] cm	Concentration[2] mg N/l	Loading, kg N/ha	Loading, metric tons of N for whole watershed
January	4.3	1.44 ± .18[3]	0.49 + 0.08[3]	16 ± 3[3]
February	4.5	1.53	0.69 ± 0.08	22 ± 3
March	10.0	1.44	1.44 ± 0.18	47 ± 6
April	9.0	1.22	1.10 ± 0.16	36 ± 5
May	5.2	0.90	0.47 ± 0.09	15 ± 3
June	3.3	0.58	0.19 ± 0.09	6.2 ± 3
July	1.4	0.36	0.05 ± 0.02	1.6 ± 1
August	0.9	0.27	0.02 ± 0.02	0.6 ± 1
September	1.0	0.36	0.04 ± 0.02	1.3 ± 1
October	1.8	0.58	0.10 ± 0.04	3.3 ± 1
November	3.6	0.90	0.32 ± 0.06	10 ± 3
December	4.8	1.22	0.59 ± 0.09	19 ± 3
TOTAL	49.7	—	5.50 ± 0.93	179 ± 30

[1] Average, 1943–1973.

[2] Based on regression NO_3-N = 0.90 + 0.63 [Sin $2\pi(m/12)$] where m = 0 for November, 1 for December, 2 for January . . . 11 for October.

[3] Calculated standard error based on variance of a predicted value of concentration and flow measured without error.

Table 3.21. Nitrate loading for several subwatersheds in the Fall Creek basin.

Watershed No.	Area km²	Number of		Regression Parameters[1]			NO₃-N Loading[2]	
		Cows	Houses	a	b	r	kg N/ha	metric tons of N, whole watershed
24	2.2	0	0	0.21	0.14	0.62	1.0 ± 0.2[3]	0.25 ± 0.05[3]
6	6.7	100	21	0.31	0.06	0.25	1.5 ± 0.2	1.0 ± 0.14
21	26.8	350	192	0.54	0.38	0.85	3.4 ± 0.9	9.1 ± 2.4
4	49.3	1175	317	1.40	0.21	0.53	7.7 ± 1.4	41 ± 8
2	76.5	—	—	1.02	0.28	0.69	6.6 ± 1.2	55 ± 10
3	14.0	387	170	0.69	0.53	0.84	4.4 ± 1.3	6.5 ± 1.9
16	107.0	2099	—	1.05	0.40	0.84	5.8 ± 1.0	60 ± 10
15	147.0	4332	—	1.22	0.70	0.83	7.3 ± 1.8	107 ± 26
1	327.0	6800	—	0.90	0.63	0.94	5.5 ± 0.9	179 ± 30

[1] NO₃-N, mg/l = $a + b [\text{Sin } 2\pi(m/12)]$, m = 0 for November, m = 1 for December ... m = 11 for October, r is correlation coefficient.

[2] Loading is sum of monthly loading. Monthly loading calculated as product of concentration estimated from regression equations and average monthly flow (per unit surface area) at Forest Home, 1943-1972.

[3] Standard error calculated from error based on variance of regression equation and assuming flow measured without error.

The NH$_4$-N concentrations were usually less than 0.2 mg NH$_4$-N/l. Thus the NO$_3$-N represents the major share of dissolved inorganic N leaving the watershed.

Sources of NO$_3$

Precipitation

Likens (1972) reports values for NO$_3$ and NH$_4$ in precipitation at several locations in central New York for the period August 1970 through July 1971. Averaged over all locations, the loading amounted to 8.8 kg N/ha. Taking this as an estimate of average annual inputs, and comparing with the results in Table 3.21, in no subwatershed did output exceed input from precipitation; that is, the watersheds varied in their ability to remove the nitrogen added in precipitation, but in no case did net output exceed the estimated inputs from precipitation.

Effects of Human Activities

The average output is a composite of many different inputs; some areas probably added large amounts of NO$_3$ while other areas added no more than was typical of the watershed uninfluenced by human activity. In this section the effects of human activities will be estimated.

During the spring and summer of 1973 measurements of NO$_3$ concentrations were made in the several subwatersheds listed in Table 3.11. The correlation matrix for the nitrate concentrations (February through April) and subwatershed variables are listed in Table 3.22. Clearly the intercorrelations among what would be judged the important independent variables is so high that the correlation analysis is useless in partitioning sources of NO$_3$ among the various human activities.

The corn land and domestic sewage were judged to be the most

Table 3.22. Correlation matrix for NO$_3$ concentrations and watershed variables.

	Area	Cows	Pasture	Corn	Alfalfa	Houses	MRP
NO$_3$ (mg/l)	.69	.76	.67	.80	.75	.82	.11
Area (ha)	1.00	.96	.96	.97	.95	.96	.20
Cows	.96	1.00	.94	.99	.99	.95	—.10
Pasture (ha)	.96	.95	1.00	.93	.96	.89	—.15
Corn (ha)	.97	.99	.93	1.00	.98	.98	—.12
Alfalfa (ha)	.95	.99	.96	.98	1.00	.93	—.17
Houses	.96	.95	.89	.98	.93	1.00	—.17

likely major sources of NO_3. The justification for this hypothesis
follows. Lysimeter data, stream data, and general agronomic information support the hypothesis that no appreciable amount of inorganic nitrogen will accumulate in soils growing grass, alfalfa, pasture, brush or forest unless they are heavily fertilized or manured.
There is no evidence that in the Fall Creek basin any of these areas
(except perhaps lawns and golf courses) are so treated. On the other
hand, the land devoted to corn receives about 90% of the fertilizer
nitrogen and 75% of the manure. In addition the corn land is often
in rotations with legume hay and during the corn portion of the rotation, the organic residues which normally accumulate under legume
hay decompose and release mineral N.

Table 3.23. Nitrogen balance sheet for 3100 ha of corn in the Fall Creek watershed, 1974. All numbers are metric tons of N.

Inorganic N	
Fertilizer nitrogen[1]	250
Mineralization of nitrogen from manure applied in current year[2]	180
Precipitation, mineralization of residual manure,	
legume residues, and soil organic N[3]	150
TOTAL	580
Crop removal[4]	270
Leached as NO_3[5]	110
Denitrified (by difference)	200
TOTAL	580

[1] 80 kg N/ha for 1900 ha corn silage; 85 kg N/ha for 1200 ha corn grain.

[2] 30 metric tons/ha (15 tons/A) applied to 3000 ha, 2 kg N mineralized per
metric ton in year of application.

[3] Approximately 50 kg N/ha/yr in excess of crop residue return.

[4] 1900 ha corn silage, 700 kg protein/ha; 1200 ha corn grain; 400 kg protein/ha.

[5] 180 metric tons of N total loading minus 35 metric tons biogeochemical
NO_3-N and 35 metric tons domestic sewage NO_3-N.

The amount of NO_3 from the individual home septic tank-disposal
field systems is unknown. In well drained areas, the bulk of the nitrogen in sewage may be converted to NO_3 and eventually leached
to streams. In the more poorly drained areas with fluctuating water
tables, the nitrogen is likely oxidized to nitrate during the drier
summer months and then denitrified during the wetter winter and
spring months. There is no means of estimating the relative importance of these systems. However, the limits of the possible impact
of domestic sewage can be estimated. Roughly, domestic sewage
contains 5 to 6 kg N/cap/year. For the roughly 12,000 inhabitants of

Fall Creek this aggregates to about 60 metric tons of N per year, or about one-third of the total nitrate output. Probably no more than one-half of this ends up in streams as NO_3, so probably the sewage from the human population contributes on the order of 30 to 40 metric tons of N.

Based on survey data (Chapter 6), an approximate nitrogen balance sheet for the 3100 ha of corn was constructed and is presented in Table 3.23. While the estimates of various inputs are clearly uncertain, they are consistent with field experience which has been summarized elsewhere (Bouldin and Lathwell, 1968). At any rate there is ample evidence in the table that the corn land is probably the source of at least 100 metric tons of NO_3-N in Fall Creek. The estimate of denitrification (by difference) is very uncertain, but denitrification must be appreciable because the inputs of inorganic N cannot be much smaller than those listed. Further aspects of nitrogen balance in Fall Creek will be examined in Chapter 4.

REFERENCES

Archer, R. J. and A. M. LaSala. 1968. A reconnaissance of stream sediment in the Erie-Niagara basin, New York. State of New York Conservation Department, Basin Planning Report ENBS.

Bormann, F. H., G. E. Likens, D. W. Fisher, and R. S. Pierce. 1968. Nutrient loss accelerated by clear-cutting of a forest ecosystem. Science 159:882–884.

Bouldin, D. R. and D. J. Lathwell. 1968. Behavior of soil organic nitrogen. Cornell Univ. Agr. Exp. Sta., Bull. 1023.

Bremner, J. M. 1965. Inorganic forms of nitrogen p. 1179–1237. In Black, C. A. (ed.). Methods of Soil Analysis. American Society of Agronomy, Madison, Wisc.

Chow, V. T. 1964. Handbook of applied hydrology. McGraw-Hill, New York, New York.

Edwards, W. M., E. C. Simpson, and M. H. Frear. 1972. Nutrient content of barnlot runoff water. J. Environ. Qual. 4:401–404.

Gilbert, B. K. 1967. Genessee River Basin. Appendix K. Sedimentation. U.S. Army Corps of Engineers. Buffalo, N.Y.

Likens, G. E. 1972. The chemistry of precipitation in the Finger Lakes Region. Techn. Rept. No. 50. Cornell University Water Resources and Marine Sciences Center, Ithaca, N.Y.

Likens, G. E. 1974. The runoff of water and nutrients from watersheds tributary to Cayuga lake, New York. Tech. Rept. No. 81. Cornell Univ. Water Resources and Marine Sciences Center, Ithaca, N.Y.

Menard, H. W. 1961. Some rates of regional erosion. J. Geol. 69:154–161.

Ryden, J. C., J. K. Syers and R. F. Harris. 1973. Phosphorus in runoff and streams. In Brady, N. C., ed., Advances in Agronomy 25:1–45.

Shelton, R. L., E. E. Hardy and C. P. Mead. 1968. Classification, New York State land use and natural resources inventory. Center for Aerial Photographic Studies. Cornell University, Ithaca, N.Y.

Taylor, A. W., W. M. Edwards and E. C. Simpson. 1971. Nutrients in streams draining woodland and farmland near Coshocton, Ohio. Water Resources Res. 7:81–89.

Taylor, A. W. and Kunishi, H. M. 1971. Phosphate equilibria on stream sediment and soil in a watershed draining an agricultural region. J. Agr. Food Chem. 19:827–831.

Vanoni, V. A. *et al.* 1970. Sediment sources and yields. J. of Hydraulics Div. ASCE. HY6:1283–1324.

ACKNOWLEDGMENTS

The personnel primarily responsible for preparation of Chapter 3 were D. R. Bouldin, A. H. Johnson and D. A. Lauer.

Special recognition is due Anne Hedges, E. O. Goyette, and Kelli Jones for their exceptional conscientiousness in the sample collection and analysis phase of the project. Kate and Hope Bouldin assisted with sample collection and analysis on weekends and at night.

Many aspects of the statistical analyses were discussed at length with Foster Cady and Jack Meisinger and their help was invaluable. Discussions with Richard Arnold were particularly helpful in interpreting soil classifications with respect to water movement.

The personnel of the Regional Office of the U.S. Geological Survey located in Ithaca, New York supplied discharge data and cooperated to the fullest extent in the hydrological aspects of the project.

Principal Authors

David R. Bouldin
Arthur H. Johnson
David A. Lauer

APPENDIX A

Details of Analysis of Stream Samples for Phosphorus

Particulate matter was separated by centrifuging samples at a relative centrifugal force of 35,000 times gravity in a Sorvall Superspeed Centrifuge for 30 minutes. The supernatant was clear enough that there was insignificant interference in a 5-cm pathlength cell. Processing and analysis were carried out within a few hours after the samples were removed from the stream since redistribution among various chemical forms was relatively rapid. Neither frozen or refrigerated storage prevented these rapid changes.

The molybdate reactive phosphorus (MRP) was determined in the supernatants using an isobutanol extraction procedure adopted from Wright (1959). The details follow.

The procedure entails formation of the heteropoly acid with ammonium molybdate, extraction of the complex into isobutanol to effect about a three-fold concentration, and subsequent reduction to molybdenum blue with stannous chloride.

Reagents

1. Isobutanol (2-methyl-1-propanol)
2. HCl, concentrated reagent, sp. gr 1.18.
3. Molybdate reagent. Dissolve 150 gr of $(NH_4)_6Mo_7O_{24} \cdot 4H_2O$ in concentrated HCl and make to 1 liter with concentrated HCl.
4. Concentrated $SnCl_2$ solution. Dissolve 20 gr $SnCl_2 \cdot 2H_2O$ in 50 ml of concentrated HCl. Store in a brown glass stoppered bottle.
5. Dilute $SnCl_2$ solution. Mix together 1 ml of concentrated $SnCl_2$ solution, 10 ml of concentrated HCl and 100 ml of absolute ethanol.

Place a 35-ml aliquot of supernatant in a 25×100 mm test tube that has a teflon lined, screw cap. Add 2.2 ml of concentrated HCl and mix. Add 12.8 ml of isobutanol, then 2.2 ml of molybdate reagent. Shake end over end for one minute and allow the phases to separate. Pipette 7.5 ml of the isobutanol into a suitable container and add 0.5 ml of dilute $SnCl_2$ solution. The $SnCl_2$ solution must be mixed fresh, and is stable for only 10–15 minutes. Mix well and allow color to develop for 30 minutes. The color is stable for at least 24 hours. Measure transmittance at 675 nanometers in a cell with a 5-cm pathlength, using pure isobutanol as a blank and carrying standards through the procedure. The method is most useful in the range 2–120 μg P per liter. Addition of HCl prior to the addition of molybdate reagent is to prevent formation of molybdosilicic acid, and the procedure will tolerate Si up to about 50 ppm.

Wright's procedure was compared to an ascorbic acid-molybdenum blue method which is essentially the single reagent method of Murphy and Riley (1962). This method extracts the molybdenum blue complex into isobutanol to concentrate the color. Transmittance is measured at 805 nm using a 5-cm pathlength. A comparison of 21 samples in the range 0–115 micrograms P/liter indicated no significant difference between the two methods at the 5% level.

The analytical procedure may be expedited somewhat by carrying out Wright's method through the extraction step and storing the isobutanol phase which may then be kept for at least a week prior to reduction and colorimetric analysis. There are slight day to day changes in transmittance upon aging, but if standards used to determine the standard curve have been carried through the storage period, accurate results are obtained. The first few steps are not time consuming and this method is further facilitated by the use of automatic pipettes and calibrated test tubes.

A 10-ml aliquot of the supernatant was transferred to a test tube and stored in these tubes until a convenient number were accumulated. All forms of phosphorus were converted to inorganic, orthophosphate form by carrying out the persulfate oxidation procedure of Menzel and Corwin (1965) in the test tubes in which the sample was stored. The inorganic phosphorus content was then determined using the procedure of Fiske and Sudbarrow (1925) in an autoanalyser.

The total phosphorus content of selected samples of the suspended solids carried in the stream was determined using the $Mg(NO_3)_2$ digestion procedure given by the AOAC (1955). The phosphorus content of the extract was determined according to the procedure of Wright described above.

References

Association of Official Agricultural Chemists. 1955. Methods of Analysis. 8th ed. p. 35.

Fiske, C. H. and Y. Sudbarrow. 1925. The colorimetric determination of phosphorus. J. Biol. Chem. 66:375–400.

Menzel, D. W. and N. Corwin. 1965. The measurement of total phosphorus in seawater based on liberation of organically bound fractions by persulfate oxidation. Limn. and Ocean. 10:280–282.

Murphy, J. and J. P. Riley. 1962. A modified single solution method for the determination of phosphate in natural waters. Anal. Chem. Acta. 27:31–36.1.

Wright, B. C. 1959. Investigations of phosphate reaction products in acid soils by application of solubility criteria. Ph.D thesis. Cornell University Library. Ithaca, New York.

APPENDIX B

Hydrograph Analysis

The empirical method of Chow (1964) was used to analyze storm runoff hydrographs to estimate surface runoff, interflow and baseflow components. Each component is considered to behave as a reservoir which fills at a rate dependent upon precipitation and infiltration and empties at a rate dependent on the head and nature of the medium. As each reservoir empties, the logarithm of the rate of outflow decreases linearly due to the decrease in head. This is generally observed in the case of baseflow and inferred for the other components. The slope of that relationship determines the recession constant for each reservoir, which is the time required for discharge rate to decrease one order of magnitude.

Figure B-1 illustrates the components of two hydrographs of storms that occurred during the experimental period. Data were col-

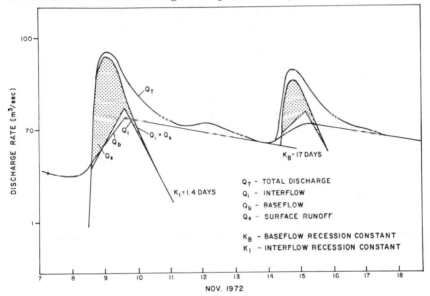

Figure B-1. An example of decomposition of a hydrograph into surface and subsurface flow components.

lected at the USGS gaging station near site 1. Details of the procedure are given by Chow (1964) and involve determination of recession constants for interflow and baseflow by inspection of the hydrographs. These recession constants ($K_{baseflow} = 17$ days, $K_{interflow} = 1.4$ days) were found to be fairly reproducible from storm to storm throughout the observation period.

The method is applicable to storm hydrographs that have one peak, and, in determining the surface runoff for the whole experimental period, estimates based on this method had to be made for multi-peaked storms.

In general, surface runoff occurred only at discharge rates above about 15 m³/sec at site 1 according to the hydrograph separation procedure. Exceptions include particularly violent storms that occurred during the summer and some storms occurring when the ground was frozen in the winter. Surface runoff as overland flow was seldom observed in the field. On these few occasions, it occurred mostly as small rivulets draining ponded water in cultivated fields that fed road ditches and eventually streams.

References

Chow, V. T. 1964. Handbook of Applied Hydrology. McGraw-Hill, New York, p. 58.

4

Flows of Nitrogen and Phosphorus on Land

Chapter 4

FLOWS OF NITROGEN AND PHOSPHORUS ON LAND

Ever since man ceased to be a roving hunter and probably even
before he took up settled life in communities, he has altered the
landscape by clearing trees, burning and cultivating, and by piling
up middens—all activities which lead inevitably to alterations in
streams.

—H. B. N. Hynes.

Models are to be used but not to be believed.
H. Theil.

INTRODUCTION

Approximately 55% of the land surface in the U.S. is used in some
way to produce food and fiber. Farming practices on this area inevitably affect and change the soil, the water flowing over and
through it, and even the air above it. A main concern of this chapter is the relation between management of farming operations and
translocation of nitrogen and phosphorus from farmed land. These
losses may be undesirable for agricultural production and deleterious for the environment. With appropriate management the
magnitude of the losses can be reduced. However, it is naive to
suppose that losses of nitrogen or phosphorus can be eliminated.
The movement of air and water inevitably removes substances
from land, with or without man's intervention.

This chapter first outlines the general characteristics of nitrogen
and phosphorus with respect to the soil–plant system, and briefly
discusses overall nitrogen and phosphorus budgets. Following this
introduction, two specific investigations are described, illustrating
the removal of nitrogen from land, in one case by air, and in the
other by water. Studies of the flow of nitrogen and phosphorus on
or from soil using detailed mathematical models are also described.

These models considered the management of nutrients designed to minimize the losses, while evaluating the corresponding effect on crop yield, and consequent return. As is obvious, there are no simple relations between the nutrients supplied and those utilized by plants. In the following chapter, economic aspects of nutrient management are further explored.

NITROGEN AND PHOSPHORUS
IN THE SOIL–PLANT SYSTEM

In managing nutrients such as nitrogen and phosphorus on land, the underlying processes governing the flows and transformations of the substances should be understood and observed. The processes relating to nitrogen and phosphorus are different. For example, nitrogen as nitrate is highly soluble and therefore subject to loss by leaching. On the other hand, only small amounts of phosphorus are lost by leaching. Another example is that while gaseous losses of nitrogen gas ammonia and nitrogen gas can occur, volatilization of phosphorus does not occur naturally. The difference between the transport of nitrogen and phosphorus in the environment is due to the latter being largely bound to the soil. This fundamental difference between the two nutrients will be stressed in Chapter 8.

Nitrogen

The general budget of nitrogen in the soil-plant system is depicted in Figure 4.1. Manure and other biological residues contain nitrogen mostly in organic form or as ammonium. Nitrogen in fertilizer may be in the ammonia, ammonium, nitrate or urea form. Ammonia and urea react with water (hydrolysis) to form ammonium, which, under aerobic conditions, is microbially oxidized to nitrate. Another source of nitrogen to the soil is biological transformation (fixation) of nitrogen gas from the atmosphere to organic form. Precipitation contains inorganic nitrogen and is therefore a further source.

Nitrogen in the soil at any particular time consists of inorganic and organic fractions. Generally, the organic pool of nitrogen is much the larger and contains components differing in biodegradability. Transformation from one form to another proceeds continuously as a result of biochemical reactions. Plants absorb primarily inorganic nitrogen and fertilizer is added to supplement the

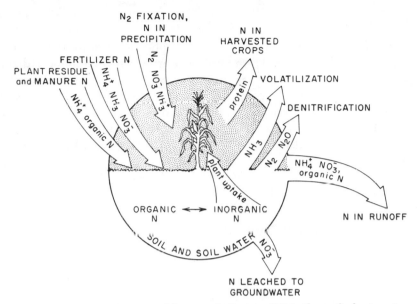

Figure 4.1. Inputs, outputs, and losses of nitrogen (N) in the soil-plant system.

existing nitrogen in the soil, especially when the rate of conversion of organic to inorganic nitrogen within the soil is insufficient for the plants.

If the application of nitrogen is mistimed or is excessive, some may be lost from the soil before the plants are able to assimilate it. The principal losses occur through denitrification, ammonia volatilization and leaching of nitrate. Denitrification is the microbial reduction of nitrate to nitrous oxide or free nitrogen which return to the air. The gases are nontoxic and, as will be described in Chapter 6, the denitrification process provides a means of removing nitrogen from waste in an innocuous manner.

Under natural conditions, ammonia does not occur in large quantities in the soil. However, in certain cases, such as application of urea and manure to the soil surface, free ammonia can be produced in significant amounts, which may escape to the atmosphere. The quantity and fate of this ammonia is largely unknown, but it probably dissolves in rainwater whereby it is returned to land or water surfaces.

Possibly a more undesirable loss is nitrate nitrogen which being highly soluble, is readily removed by water flowing over or through soil. Not only can this represent an economic loss in crop production, but the nitrate may, in excess, contaminate water resources.

Phosphorus

Plants require considerably less phosphorus than they do nitrogen. However, the proportion of phosphorus to nitrogen in fertilizer is often greater than the corresponding proportion found in plants. This is because soluble forms of phosphorus react rapidly with soil to form slightly soluble inorganic compounds, which are sparingly available to plants. They are also relatively immobile and hence inorganic phosphorus is, in certain important respects, the antithesis of inorganic nitrogen.

The generalized movement of phosphorus in the biosphere is represented in Figure 4.2. The organic forms are probably less im-

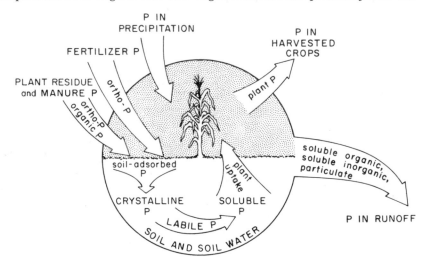

Figure 4.2. Inputs, outputs, and losses of phosphorus (P) in the soil-plant system.

portant than inorganic forms to plants in most soils. As plant roots remove phosphorus from the soil solution, phosphorus adsorbed to soil particles enters into solution to at least partially replenish that removed. This replenishing or labile pool of phosphorus constitutes only a small fraction of the total in the soil.

As shown in Figure 4.2, the removal of phosphorus from the soil is almost entirely due to plants and water. Gaseous losses of phosphorus do not occur naturally. However, doubtless some phosphorus becomes airborne in dust.

Generally, water in contact with unfertilized soils and the unconsolidated mantle contain relatively low concentrations of phosphorus. Applications of manure and fertilizer to soil can change the concentration of nitrogen and phosphorus in the soil water. It follows that surface runoff from manured and fertilized soils is more likely to contain higher levels of phosphorus than subsurface flow, since fertilizers and manures are generally applied to the plow layer.

GENERAL BUDGET OF NITROGEN AND PHOSPHORUS IN HUMAN FOOD PRODUCTION

The nitrogen and phosphorus in human food production is approximately quantified in a generalized food web shown in Figure 4.3, where the inputs and outputs of the nutrients are expressed as kg per capita per year. It is evident from the food web depicted that inputs exceed the output in food by a considerable margin. For example, the diet of each person contains an average of about 6 kg N and 0.6 kg P per year, but inputs to the soil-crop system are approximately 50 kg nitrogen and 12 kg phosphorus. Approximately 2/3 of the nitrogen consumed by man is in the form of animal protein. (Other goods such as cotton and wool, produced on farms in the U.S. utilize relatively small amounts of nitrogen and phosphorus compared to food).

Animals convert only a relatively small fraction of plant to animal protein. For every 10 kg N and 20 kg P in animal feeds, only about 1 kg of each eventually is available in food consumed by humans. The amounts not accounted for in output as food are excreted and either recycled or otherwise lost from farming. As shown in Figure 4.3, animal nutrients are usually obtained from cultivated crops, pasture, rangeland and manufactured feed supplements. Quantification of the different components on a national scale is virtually impossible.

Quantification of all the sources of nitrogen and phosphorus for crops is equally difficult. Fertilizers are a major source, with about 35 kg and 9 kg of nitrogen and phosphorus respectively being applied in fertilizers per capita per year. The corresponding inputs from precipitation, the soil and manure are much less easily ascertained. The amount of nutrients contributed from animal manure is unknown because of unmeasured losses between excretion by the animal and application to land. To determine these losses nationally is impossible because they depend on many highly variable factors especially those governing the treatment, storage and disposal of the manure. [See manure handling in Chapter 5 and the control of nitrogen in manures in Chapter 6.]

The net addition to the human food web is estimated as 50 kg

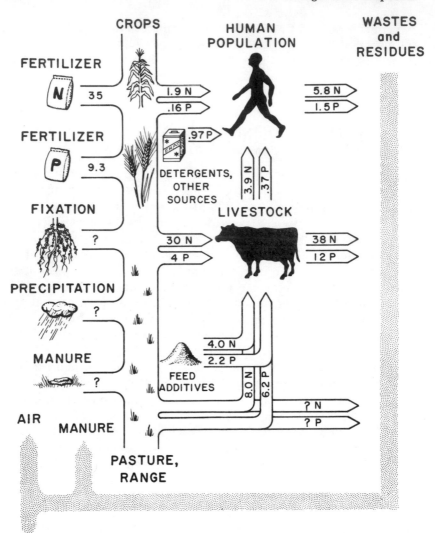

Figure 4.3. Nitrogen and phosphorus flow in the human food web in the United
States (kg/capita/yr).

N/cap/yr (35 kg N as fertilizer, 15 kg N from fixation). The difference between this amount and the 6 kg N/cap/yr in the human diet,
44 kg N/cap/yr, represents nitrogen which may be removed by runoff, leaching of NO_3, denitrification, NH_3 volatilization, or stored
within the soil. In later sections the loss of NH_3 by volatilization
from animal manures will be discussed and its significance assessed.

Denitrification and leaching are so variable from area to area that any generalization nationally would be questionable.

With respect to phosphorus, the major sources for plants are fertilizer, some unknown quantity provided from the soil, plant residues and animal wastes. The amount of phosphorus in domestic sewage (Figure 4.3) is greater than that which would be predicted when dietary intake alone is considered. This is due to contribu-

Figure 4.4. The application of fertilizer in early spring places the nutrients in the ground before the seeds have germinated.

tions to household wastes from phosphorus added to water supplies for the control of corrosion in pipes, various cleansing products, unused food washed down drains and especially to the use of phosphatic detergents. When the latter source was abolished by law in New York State on June 1, 1973, the per capita discharge of phosphorus to sewers decreased by about 50%. (Valentine, 1974)

Summary of Nitrogen Budget in Fall Creek

Investigations described in this book produced a great deal of information about the use of nitrogen in the Fall Creek watershed. This information was used to construct a budget for the nutrient as described below. While much of the data have considerable uncer-

Figure 4.5. Thunderstorm over Lake Cayuga: a potential source of nitrogen for
the soil and water.

tainties, it is unlikely these are sufficient to negate the conclusion
that large amounts of nitrogen cannot be accounted for.

A summary of the data is shown in Table 4.1(a). A main aim of
this table is to provide the estimate of the nitrogen which is unaccounted for and presumably stored in the soil, or lost through
utilization of ammonia, or through denitrification. This estimate is
calculated as the difference of the sum of inputs and the sum of
outputs, of nitrogen to and from the watershed.

Inputs and outputs of nitrogen are summarized in Table 4.1(a)
under the headings "crops", "animals" and "other." The estimated
annual amount of nitrogen potentially available for input to crops
is approximately 1800 metric tons, comprised of new and recycled
inputs. The "new" inputs (that is not recycled) are precipitation,
fertilizer, and symbiotic fixation, which are shown at the top of
Table 4.1(a) between the dashed lines (as are the other "new" inputs). The precipitation was obtained from Likens (1972), and the
fertilizer input from Chapter 5. Leguminous fixation was estimated

Table 4.1(a). Summary of estimated nitrogen balance in Fall Creek watershed (metric tons).

Type	Crops	Animals	Other	Summary of Balance
New Inputs	Inputs Precipitation 140 Fertilizer 270 Leguminous fixation 260 670	Inputs Purchased feed 320 → 320	Inputs Food for population 85 Precipitation on non cropland 165 → 250	Inputs Sum of inputs to the watershed →1240
Recycled Inputs	→ Manure 960	→ Feed in watershed 880		
Sum of Inputs	1630	1200	250	1240
Outputs	Outputs Corn (silage and grain) 270 Legume hay 340 Grass hay 190 Pasture 80 880 →	Outputs Milk 190 Meat 50 240 960[1]	Outputs (Fall Creek) Nitrate-nitrogen 180 Particulate-nitrogen 50 → 230	Outputs Sum of outputs from the watershed → 470
Unaccounted for	← 750[1]		→ 20[1]	→ 770[1]

[1] Differences between summed inputs and summed outputs.

to be half the total nitrogen in the harvested hay crop (grass and legume hay; the remainder of the nitrogen in the hay was postulated as being derived from residual manure and legume residues). The recycled nitrogen which is potentially available for crops includes 960 metric tons of nitrogen in manure. Allocating part of this nitrogen elsewhere in the table would alter the ratio of inputs to output for crops, but would not affect the final sum as described above. According to the table, more than half the inputs to crops, 880 metric tons, is eventually consumed by animals, a ratio which commonly occurs. The unaccounted for nitrogen under the heading "crops" is 750 metric tons.

The total input to animals is estimated at 1200 metric tons of nitrogen of which 880 metric tons are from feed grown in the watershed and 320 metric tons of "new" nitrogen or in the form of purchased feed originating outside the watershed (Chapter 5). Of this total, 190 metric tons of nitrogen are converted to milk and 50 metric tons to meat, yielding a total output of 240 metric tons. It may be assumed that most of these products are consumed outside the watershed and hence represent exports from the watershed.

The difference between inputs in feed and outputs in milk and meat is 960 metric tons which is manure nitrogen and is available for return to the crop system as pointed out above.

Finally additional "new" inputs summarized under the heading "other" includes inputs of food for the human population and precipitation on other than agricultural land. Human food contains about 6 kg N/cap/yr as shown in Table 4.1(a), and it is assumed that the total consumed by the population is excreted. Sawyer (1965) has estimated that an average man consumes in his lifetime 380 kg N of which 99.5% is excreted.

The "other" outputs include 180 metric tons of nitrate-nitrogen in Fall Creek and 50 metric tons of particulate nitrogen: giving a combined total of 230 metric tons which leaves the watershed via the Creek.

The greatest interest lies in the unaccounted for 770 metric tons of nitrogen. This compares to an approximate total input of "new" nitrogen to the watershed of 1230 metric tons. Probably the nitrogen content of the farmed soils is very nearly in "steady state", that is, inputs equal outputs. Thus most of the 770 metric tons of nitrogen which are unaccounted for probably return to the atmosphere as NH_3, N_2 and N_2O. Data are not available for estimating what fraction is lost in each form, although as will be documented later, ammonia volatilization probably accounts for a substantial fraction.

In summary, as may be seen from Table 4.1(a), the approximate total nitrogen "lost" from agriculture is 750 metric tons plus the 230 metric tons conveyed in Fall Creek, giving about 1000 metric tons. This represents about 60 kg/ha if the losses were entirely attributed to farmland.* This raises two questions: a) could farmers benefit financially from reduced losses, and b) what are the effects on the environment. As already argued in Chapter 3, the levels of nitrogen in Fall Creek do not present any well defined environmental hazard. With respect to the nitrogen lost to the atmosphere, further discussion can be found in Chapter 8.

Summary of Phosphorus Budget in Fall Creek

Table 4.1(b) was constructed to represent the annual budget for phosphorus in the Fall Creek watershed. The numbers were generally estimated using the sources as described for the nitrogen budget, except for the assumed value for the phosphorus content of steer carcass of 0.74% (Maynard and Loosli, 1969).

Total new inputs of phosphorus to the watershed were estimated to be 237 metric tons annually. Outputs in farm produce were estimated to be 38 metric tons per year, which when summed with that transported out of the watershed by Fall Creek, gives a total amount of 58 metric tons, which leaves the watershed annually, a ratio of input to output of 4:1. In the case of nitrogen it was argued that most of the unaccounted for difference between inputs and outputs was presumably "lost" to the atmosphere, whereas it is probable that most of the 179 metric tons of phosphorus, which are unaccounted for, are retained in the watershed.

It may also be noted that, based on the number derived in Table 4.1(b), the amount of biologically available phosphorus transported by Fall Creek is less than 3% of the phosphorus estimated as entering the watershed each year. This suggests that changes in the levels of inputs of phosphorus, unaccompanied by changes within the watershed itself, are unlikely to cause substantial changes in the amount of biologically available phosphorus leaving the watershed via Fall Creek.

---

* Some of the nitrogen in the creek originates from nonagriculture; hence in this sense, the rate of loss of nitrogen from farmland is overestimated. However, the total amount of nitrogen carried in Fall Creek exceeds the 240 metric tons estimated as leaving the watershed in the river, because of losses of nitrogen within it. To some undetermined extent, these "errors" may balance each other.

Table 4.1(b). Summary of estimated annual phosphorus balance in Fall Creek watershed (metric tons)

Type	Crops	Animals	Other	Summary of Balance
New Inputs	Inputs	Inputs	Inputs	Inputs
	Precipitation 3	Purchased feed 96	Population[2] 21	Sum of inputs to the watershed
	Fertilizer 114		Precipitation on non cropland 3	
	→ 117	→ 96	→ 24	→ 237
Recycled Inputs	⌐ Manure 168	→ Feed in watershed 110		
Sum of Inputs	285	206		237
Outputs	Outputs	Outputs	Outputs (Fall Creek)	Outputs
	Corn (silage and grain) 49	Milk 31	Total soluble phosphorus 5	Sum of outputs from the watershed
	Legume hay 31	Meat 7	Particulate phosphorus 15	
	Grass hay 26			
	Pasture 4			
	110 →	38	→ 20	→ 58
Unaccounted for	175[1]	168[1]	→ 4[1]	→ 179[1]

[1] Difference between summed inputs and summed outputs.
[2] Includes primarily food and detergents.

It should be noted that the nutrient budgets for Fall Creek represent an ambitious attempt to address a difficult problem. Only recently have ecologists begun to quantify similar budgets for much simpler systems, namely forested watersheds in various successional stages (e.g., Vitousek and Reiners, 1975 and Likens et al., 1970). By comparison with these, the Fall Creek watershed is much more structurally complex and has higher and more variable rates of change in nitrogen and phosphorus input and output. The variety and ubiquity of human activities, especially farming, in the watershed of Fall Creek may contribute to analytical difficulties; however these are the very features that cause it to be of interest as a prototype of diffuse nutrient sources.

MANAGEMENT OF NITROGEN ON THE FARM

The application of large amounts of nitrogen in farming in the form of animal manure and inorganic fertilizer, and the substantial fraction lost to air and water, prompts related questions: (a) How can the utilization of nitrogen in fertilizer by crops be improved? (b) What happens to nitrogen in animal manures? Partial answers to these questions are noted in the following sections.

Volatilization of Ammonia from Manure

Volatilization of ammonia could represent a large component of nitrogen flow in both the Fall Creek watershed and United States nutrient balances. Volatilization of ammonia from manure does occur. Evidence provided by the nose can be confirmed by measurement showing that the levels of free ammonia, in the immediate vicinity of animal production units, are higher than in the environs generally.

Animal wastes contain enzymes and microorganisms which rapidly transform a substantial fraction of the total nitrogen in the wastes to an ammoniacal form, and the manure begins losing ammonia almost immediately following excretion. This process continues until the ammoniacal nitrogen is exhausted or conditions favor transformation or adsorption of the ammonia such as occurs following mixing of the manure with soil.

This section describes an investigation which quantified the losses of ammonia to the air from manure following its distribution on the surface of the soil. Since dairy farming is the dominant agricultural enterprise in the Fall Creek watershed, as it is in New York

State, dairy manure was used in the studies. Given the nature of the microbial processes governing the production of ammonia, it seemed probable that the rate of volatilization would be subject to climatic conditions. Therefore, five experiments were performed in the field over two years.

Measurements of the total ammoniacal nitrogen were made using samples of manure collected directly from the soil surface. By frequent sampling following spreading of the manure, rate of loss of ammonia over time was determined. The rates of application and corresponding measurements are shown in Table 4.2.

Losses of ammonia are summarized in Table 4.3. No losses from volatilization occurred during January 1974 because of a combination of subfreezing conditions, subsequent snow and a rapid thaw, which leached the ammoniacal nitrogen into the soil. The total nitrogen applied per hectare over the entire period was 5227 kg. This included 1366 kg of ammoniacal nitrogen of which an estimated 892 kg was volatilized.

The rate of loss could be approximated by a first-order rate process (rate of loss proportional to amount present), and the parameter representing the half-life of the substance (time required for 50% of the ammoniacal nitrogen to be volatilized) was adopted to compare rates of ammonia loss. The low and high rates of application had mean half-lives of 1.86 to 3.36 days, respectively, excluding those made in January. After the initial loss, rates usually decreased. The higher rate of loss observed in the manure spread at a rate of 34 metric ton/ha is probably due to the relatively faster drying of the thinner layer of manure on the soil surface.

The results of this work underline the importance of managing manure not only in disposal, but immediately following excretion. Greatest losses of ammonia occur following defecation and the consequences of this must be assessed (Table 4.4). As discussed in Chapter 6, the management of animal wastes to control both odor and levels of nitrogen in the manure is possible.

It is concluded that volatilization of ammonia represents a large fraction of the total losses of nitrogen depicted in Figure 4.3 and Table 4.1(a). In fact, evidence presented here and in the literature suggests the hypothesis that approximately half of the nitrogen in animal manure in the United States is lost by volatilization as ammonia. This large quantity of nitrogen is approximately equal to half the amount of nitrogen in fertilizer used in farming and is three times greater than the estimated content of nitrogen in the human diet. Nationally, it is equivalent to nearly 4 kg N/ha/yr over the

Table 4.2. Rates of application of manure.

Experimental Period	Rate of Wet Manure Application metric tons/ha	Dry Matter %	% of Dry Matter	
			Total N	Ammoniacal-N
April 9–30, 1973	225	17.1	2.85	0.909
August 13–September 7, 1973	225	18.4	2.41	0.457
January 7–March 4, 1974	200	27.5	1.81	0.481
January 10–March 4, 1974	34	21.9	1.86	0.503
April 24–30, 1974	200	21.4	1.78	0.426
April 25–30, 1974	34	21.7	1.37	0.271
May 27–June 12, 1974	200	21.2	2.18	0.653
May 31–June 12, 1974	34	30.0	2.04	0.444

Table 4.3. Summary of nitrogen budget in applied manure.

Experimental Period	Rate of Manure Application metric tons/ha	Nitrogen Budget (kg N/ha)			NH₃ Lost as % of Applied Ammoniacal N
		Total N Applied	Ammonia Applied	NH₃ Lost	
April 1973	225	1100	350	285	81
August 1973	225	998	189	189	99
January 1974	34	996	265	0	0
	200	178	37	0	0
April 1974	34	762	183	17	85
	200	101	20	112	61
June 1974	34	924	277	41	91
	200	208	45	248	90
					65[1]
Totals	1152	5227	1366	892	SD 13

[1] Weighted overall mean loss with a standard deviation of 13.

Table 4.4

Manure Condition	Dry Matter %	Total N %	Ammoniacal N as % of Total N
As defecated[1]	11.4	5.65	60.7
Farm-fresh[2]	15.0	3.07	36.2
Farm-stored[3]	24.0	1.84	25.1
Farm-stored after spreading[4]	46 to 85	1.46	4.52

[1] Derived from experimental feeding trials in which urine and feces were collected and analyzed separately (Personal communication with Carl E. Coppock, Animal Sci. Dept., Cornell Univ. and Fisher (1974)).

[2] Obtained from dairy used as source of manure in this investigation. This manure was ≦24 hours old and obtained at daily cleaning time. (Collected February 29, 1972).

[3] Obtained from same dairy farm but stored in an unprotected pile for an indefinite period.

[4] This is mean condition of the "Farm-stored" manure from the 1974 series of experiments several days after spreading on the soil surface.

whole land area of the continental United States. Even assuming that this estimated loss is overestimated by 100%, it would still represent a very large part of the total nitrogen lost annually from the human food web. The implications of this loss for the quality of the atmosphere are not known. Presumably most or all of the nitrogen apparently returns via precipitation with unmeasured consequences for the ecosystem. As discussed previously there are the dual questions: What are the environmental effects of the loss, and should attempts be made to conserve the nitrogen from the viewpoint of agricultural production? (See Chapter 8.)

Improving Recovery of Fertilizer Nitrogen by Crops

Introduction

The previous discussion makes clear that nitrogen from fertilizer is of major importance in the human food web. Currently, 9 million metric tons or 35 to 40 kg N/per capita are used annually in the United States. Given this heavy consumption and its increasing expense for the farming industry, and its possible effect on the national ecosystem, it is pertinent to consider some of the issues governing the use of fertilizer.

Farmers frequently manage nitrogen additions (in the form of fertilizer, manure and leguminous residues) so that the total supply of inorganic nitrogen is in excess of the amount needed for maximum crop yields. Generally in the past, costs of fertilizer were relatively low and this practice has been economically reasonable. In effect, farmers have added fertilizer nitrogen to ensure nutrients for the crop to maintain maximum yield under all conditions. Thus, there are conditions where excess nitrogen may be leached away as nitrate or lost through denitrification.

The second reason farmers often apply more nitrogen is that it is convenient to apply fertilizer when demands on farm labor are low, which may be considerably in advance of the period of most rapid uptake by the crop. For example, in the midwest, farmers often apply nitrogen early in the spring for corn, many weeks prior to the period of the crop's maximum rate of growth. Thus, there is a long period during which nitrogen may be leached into groundwater or lost to the atmosphere through denitrification. Farmers therefore apply an excess of fertilizer in anticipation of such losses. Furthermore, they generally allow for worst situations, that is, enough fertilizer is added to maintain yields even when such losses are highest.

A third reason concerns the availability of fertilizer. In many regions of the United States (e.g., midwestern corn belt where over 50% of the fertilizer nitrogen is used), very large acreages of identical crops are grown and, if the application of fertilizer were restricted to a short period dictated by crop conditions, then a large fraction of the total fertilizer used would be applied nearly simultaneously. The system of distribution of fertilizer from the manufacturer could be overburdened if confronted with simultaneous demands. Also, the capital costs of storage for such a large single period of application could be high. It is accepted therefore that fertilizers may be applied at times inconsistent with crop requirements. Nevertheless changes in applications are possible, and perhaps desirable to conserve both energy and the environment.

To support this thesis, an investigation was made of potato production on Long Island, New York. An account of the investigation follows.

Efficiency of Fertilizer Use on Long Island Related to the Contamination of Ground Water by Nitrate

Long Island lies just east of New York City. It encompasses a main island which is 120 miles long, and several smaller ones, the total being 1400 square miles in area. Ground water aquifers are the only source of natural fresh water, and in recent years the Island has become one of the few areas in the northeastern United States where contamination of ground water by nitrate-nitrogen is a serious problem. In parts of the more rural areas of eastern Long Island, the ground water contains more than 10 mg N/liter in the form of nitrate. The degree of contamination is more apparent in this eastern area because the underground aquifers are fairly shallow. There is an accumulation of inorganic nitrogen in the ground water throughout the Island, and if unchecked, eventually the deeper and larger aquifers in the western part also will become polluted.

Accumulations of nitrate in ground water are derived from an unknown combination of fertilizer and domestic sewage. The population itself is potentially a relatively huge source of nitrogen. For example, if all the human waste were uniformly distributed over the Island, the rate of application of nitrogen would be over 100 kg N/ha/yr. Also, use of fertilizers is relatively high on the approximately 32,000 ha (or 9% of the total land surface) in agricultural production. The most eastern county, Suffolk, has the highest value of farm production in New York State, largely from potato, sod and horticultural products. The county is also one of the largest producers of ducks in the United States.

At present, the population of Suffolk County is steadily increasing with a corresponding increased burden being imposed on the Island's water resources, both through greater consumption of water and the additional volumes of wastewater that result. Individual home sewage disposal systems perform very satisfactorily in most areas of the county, but probably most of the nitrogen in the sewage (5 to 6 kg per capita per year) is ultimately converted to nitrate and leached to the ground water. If large sewage collection systems were installed and the effluent discharged into the ocean rather than recharged to the ground water, two undesirable consequences could follow. There could be a net export of water from the Island, and salt water could encroach into the depleted aquifers. Also, the resulting nutrient enrichment of the marine environs might be unacceptable.

Estimates of the amount of nitrate derived from the 10,000 hectares in potato production indicate that it may be equivalent to the nitrogen in sewage from 150,000 to 300,000 persons, *i.e.*, about 1200 metric tons per year or 100 kg/ha/yr. There are several ways in which the load could be reduced. First, potato production could be prohibited. However, this would eliminate an important economic activity. Second, the amount of fertilizer used could be limited. This latter alternative seems feasible both economically and socially, but

Figure 4.6. Intensive agricultural production in Long Island: potato crops.

field experiments are necessary to ascertain how more efficient use of fertilizer nitrogen can be accomplished without seriously reducing income to the potato farmer and his contribution to the economic viability of the area.

Presently, potato farmers are applying 200 to 250 kg N/ha in fertilizer each year. Removal of nitrogen in the potato tubers amounts to about 110 to 130 kg N/ha/yr, or approximately half of that applied. In the well-drained areas of Suffolk County where potatoes are grown, most of this nitrogen is leached to ground water as nitrate. The farmers apply most of the fertilizer nitrogen at planting time in middle to late April. The potato plants emerge about one month later, and approximately 6 weeks after planting begin to take up nitrogen rapidly. However, during the 6 weeks between planting and the beginning of the rapid uptake period, an appreciable fraction of the nitrogen may be leached as nitrate, from the rooting zone, particularly in years that have more than average precipitation before the plants emerge. The farmers apply enough fertilizer every year at planting time to supply nitrogen for a potato crop during a wet year as a hedge against the possibility that the coming season may be wetter than average. The average annual rainfall is 42 inches and hence the probability of heavy rainfall is high. If leaching does not occur during the growing season, then excess nitrogen remains in the soil as nitrate at the end of the season, and it will be leached by fall and winter rains into the ground water. Probably the present fertilizer practice very nearly maximizes returns so far as the individual farmer is concerned (ignoring the externalities of pollution of ground water).

Field tests of the hypotheses, that proper timing of fertilizer will reduce the amount needed relative to current procedures, were carried out in 1970, 1971, 1972, 1973, and 1974 on the vegetable research station at Riverhead (Long Island), New York. The experimental procedure was to apply a small amount of fertilizer at planting, followed by additional amounts soon after emergence. This procedure is termed split application. Since this reduces leaching losses during wet years, only a small allowance has to be made for such losses and hence only enough fertilizer to meet the requirements of the potato crop need be applied. Yields of potatoes with split application of lesser amounts of fertilizer should be as high as those obtained using the present method of applying 200 to 240 kg/ha of fertilizer nitrogen.

The results were as follows. Yields obtained when nitrogen applications were split between the time of planting and 6 to 8 weeks afterwards were equal or superior to those obtained when all of the nitrogen was applied at planting. This is illustrated by the 1972 and 1973 experiments summarized in Table 4.5. In 1973 a more comprehensive experiment was performed with the objective of defining the quantity of nitrogen needed for economically optimal returns. The results are shown in Table 4.6. The efficiency of nitrogen use was quite high because the amount of precipitation was not excessive enough to cause serious leaching losses even when all the nitrogen was applied at planting.

The potato harvest during the period of the investigation was about 30 metric tons/ha of tubers containing about 120 kg N/ha. This is about half the amount of nitrogen normally applied in fertilizer. However, as can be seen from Table 4.6, it may be possible to sustain similar yields with split application of fertilizer which have a small combined rate of application. Hence if yields of 30 to 35 metric tons/ha can be obtained with split applications of 125 to 150 kg N/ha, no more than 20 to 40 kg N/ha on the average will be lost to the ground water. Since this amount of nitrogen will be dissolved in approximately 50 cm of recharge water, the leachate will contain less than 10 mg N/l as nitrate (50 cm of water per hectare $= 5 \times 10^6$ liters), assuming no other sources.

The results from this investigation demonstrate that economic

Table 4.5. Yields of potatoes in 1972 and 1973 with different amount of fertilizer nitrogen and different methods of application at the Vegetable Research Station, Riverhead, New York.

Method of Application	Yield of Tubers, metric tons/ha					
	90 kg N/ha			180 kg N/ha		
	1972	1973	mean	1972	1973	mean
All at planting[1]	21	28	24	23	33	28
Split[2]	22	31	27	26	32	29

[1] Average of broadcast and band methods of application.
[2] Average of two methods of splitting:
  Method A: 1/2 at planting, 1/4 at emergence, 1/4 at 6"–8" height
  Method B: 1/2 at planting, 1/4 at 6"–8" height, 1/4 during last week of June.

yields of potatoes can be obtained consistent with acceptable nitrate loss to ground water. In effect, less fertilizer will produce the same yield, when the application is properly timed, as some larger amount applied at a less favorable time. The cost is usually not excessive and may even reduce production costs slightly. The farmer spends more money, time and management skill to apply the fertilizer at the proper time (usually this requires an extra trip through the field), than would be required if all were applied at planting or at some other time, but these costs are partially or completely offset by the purchase of less fertilizer.

Currently the Cornell University Experiment Station and Suffolk County are carrying out an extension-demonstration-research program on Long Island with the objective of persuading the farmers to apply fertilizer in amounts and timing that correspond more closely to plant requirements. There is thus a keen interest on the part of local government officials to seek ways of preserving the

Table 4.6. Yields of potatoes in 1973 with different quantities of nitrogen applied in split application on the Vegetable Research Station, Riverhead, New York.

Quantity of Fertilizer Nitrogen Applied, kg N/ha	Yield of Tuber, metric tons/ha	
	Split[1]	Band at Planting
0	19	—
78	30	—
112	31	31
168	31	31
224	—	31

[1] 56 kg N/ha at planting, remainder applied at emergence.

Island's water resources. Indications are that within the next few years most farmers will apply less fertilizer in such a way that it is used more efficiently with reduced losses by leaching. As previously explained, this will reduce the total amount of nitrogen applied without diminishing crop yields. Another study is underway to investigate the viability of a similar strategy in turf management.

Reference to Other Investigations on Timing of Fertilizer Application

Although only a passing reference can be made, there are many other relevant studies on the timing of applications of nitrogenous fertilizers which have been reported. Similar experimental comparisons of fertilizer applied to corn in fall, spring and as a summer sidedressing have been performed (i.e., Lathwell, Bouldin, Reid, 1970; Stevenson and Baldwin, 1969). Although these investigations were carried out in a wide range of soil and climatic conditions, the results are remarkably similar and substantiate the findings of the experiments on Long Island as previously summarized. In addition to corn and potatoes, crops such as cotton, winter wheat, other small grains and certain grass crops respond similarly to time of nitrogen application.

The full economic advantages and disadvantages of sidedressings, or split applications of fertilizer, depend upon the value of the crop relative to the cost of nitrogen, labor costs and availability, distribution and storage facilities for fertilizer, and the external costs of nitrate pollution. A major difficulty in such a determination is the probabilistic nature of crop yields and nutrient uptake and losses especially as these respond to rainfall and temperature. Assuming economic feasibility then, most studies clearly support the policy of applying fertilizer just prior to the period of maximum crop growth.

MATHEMATICAL MODELS FOR NUTRIENT MANAGEMENT

The investigations described in previous sections illustrated two routes by which nitrogen could be lost from farmland following the application of manure and fertilizer to the soil. In the following section, studies based on the use of large-scale mathematical models are briefly outlined. These attempted to comprehensively account for the inputs and losses of nitrogen, and to a lesser extent phosphorus, to and from the soil.

Introduction

For immediate purposes, mathematical models will be taken to be large-scale representations of a system or subsystem, requiring the use of a computer. The construction and use of such models provide two major advantages. First, they can be an aid for planning or management purposes. Although management can obviously be formulated without mathematical models, these are able to accept a huge amount of data and perform complex operations on it in accordance with scientific knowledge. The consequences of any specified assumptions within the restrictions of the data can thereby be computed. In addition, scientific research itself may be promoted by the use of the model providing insight into existing investigations, or indicating new areas requiring research.

All mathematical models are idealizations, however, and their representations depart, to a varying extent, from the natural phenomena being modeled. These departures can be cumulative, and therefore serious in very large models. While part of the model may not be seriously inaccurate, combined errors in the whole may be. Models are limited by computing capacity and available data and are, therefore, a combination of generalization and omission, constructed hopefully without unduly sacrificing their reliability.

It may be argued that scientific knowledge is inadequate and therefore comprehensive models will be inescapably wrong. This argument, in terms of absolute truth, is irrefutable. However, the acceptance of any model is discretionary, and it should be possible to interpret conclusions, given a clear understanding of the assumptions and data used.

Cornell Models for Nutrient Management

Of four large-scale models constructed at Cornell, the first two described were concerned solely with the management of nitrogen. The remaining two considered both nitrogen and phosphorus. Three of the models also evaluated economic considerations of assumed management constraints. All four models considered some aspects of the flow of water through the watershed, the movement of sediment and nutrient flows together with their associated transformation, transport, and inventories.

A simple representation of nitrogen flow in a watershed is shown in Figure 4.7. Transformation, transport, and storage of nitrogen are indicated by circles, triangles, and squares, respectively. A corresponding flow diagram is reproduced for each of the models concerned with nitrogen, providing a visual comparison of each treatment of the flow.

General Observations of the Four Models

A general indication of the main characteristics of each model is presented in Table 4.7. An outline of the transformations assumed

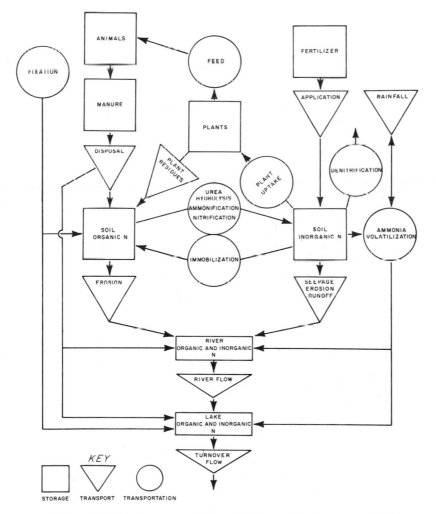

Figure 4.7. Simple representation of nitrogen flow in a watershed.

Table 4.7. General summary and comparison of large-scale Cornell models for nutrient management.

Parameter	Major Investigator			
	Haith [1] Model A	Schulte [2] Model B	Coote [3] Model C	Schaffer [4] Model D
Pollutant	N	N	N & P[1]	N & P[1]
Scope	farm	farm	farm	watershed
Emphasis	simplified soil N cycle and land disposal policy	N cycle of an egg production facility and detailed soil moisture balance	soil characteristics, simplified nutrient budgets and legislation impacts on dairy farms	detailed soil N cycle and economic analysis of nutrient control policies
Time scale	monthly	daily	seasonal	daily & yearly
Model type	analytic & L.P.	event simulation	L.P.	event simulation and L.P.
Hydrology	simplified monthly soil moisture balance with the assumption of no runoff	modification of USDAHL–70 watershed model of hydrology	Soil Conservation Service runoff equation	modification of Hahm's small watershed model
Sediment	N.A.	use of the Universal Soil Loss Equation on an event basis	use of the Universal Soil Loss Equation on a seasonal basis	use of Meyer and Wischmeier soil loss model on an event basis
Waste management	storage and disposal	storage, treatment and disposal	storage and disposal	—
Economics	aggregated costs and income with assumed constraints	N.A.	detailed costs and income with assumed constraints	aggregated costs and income with assumed constraints and with tax constraints
Data requirement	small	large	very large	very large

1 P adsorbed on sediment.

in the nitrogen flow of the models is given in Table 4.8. These tables summarize the main common features and distinctions between the models.

Although there are differences in objectives and techniques used by the models, they have one overriding principle in common. That is the manner in which the watershed, or farm unit, is separated mathematically into its constituent parts: water flow, soil loss, nutrient flow and economic factors. These parts are then reintegrated by the appropriate sequence of superimpositions.

Table 4.8. Aspects of the nitrogen flow represented by the Cornell models.

	Model			
Aspect	A	B	C	D
Urea hydrolysis				*
Ammonification	net	net	net	
Immobilization	mineralization	mineralization	mineralization	*
Nitrification				
Denitrification				*
Seepage losses	*	*	aggregated	
				*
Runoff losses		*	*	*
Volatilization			*	*
Plant uptake	*	*	*	*

* Inclusion of the aspect of nitrogen flow in the model.

To produce a mathematical structure which reasonably depicts reality, the models are dependent on field data and the results of basic scientific research. Where there are inadequacies in one or both of these, the models are accordingly less useful. An example common to three of the models is the use of the universal soil loss equation. This requires not only a large amount of data for its application, but fails to produce good estimates of the total sediment eroded from a watershed and deposited into streams and lakes. The mathematical form of the equation needs modification by incorporating a transport relation which adequately predicts deposition of some sediment before it reaches a waterway.

The problem of applying the soil loss equation to a large area is a particular example of a fundamental distinction between the models. That is the time and spatial scales they assume. Data requirements and even the mathematical assumptions are dictated by the choice of time or spatial scale. A larger time scale permits an aggregation of data to represent average conditions, and as a result single coefficients, or equations, may be used to represent several processes. An example of this is shown in Table 4.8. Models A, B and C all considered the net result of mineralization of organic nitrogen rather than the individual transformations involved. This aggregation can be especially desirable if the individual processes are not fully understood or if data are inadequate. Aggregation of variables with respect to space, however, may be less successful because the averaging process may compound rather than cancel errors; a possibility which will be referred to in the discussion of the models.

REVIEW OF MATHEMATICAL MODELS FOR MANAGEMENT OF NUTRIENTS IN WATERSHEDS

Model A: Optimum Control of Nitrogen Loss from Land Following Disposal of Organic Wastes (Haith, 1973)

Description of Model A

Model A primarily evaluated the economic and environmental consequences of disposing nitrogenous wastes on cropland. A linear programming model was used to estimate maximum profit in a hypothetical farming unit, given assumed constraints on losses of nitrogen to ground water. The management decisions examined were the size of the area on which the wastes were applied, storage capacity for the wastes prior to disposal, and the volume of waste disposed on the land per month. Although the model specifically considered sewage sludge and dairy manure, it was formulated so the disposal of other wastes could be assessed.

The basis of the model was a simplified nitrogen flow into and out of the soil (Figure 4.8). Linear equations were used to represent the additions and losses of the nutrient to the 'soil bank'. Two equations calculated for each month describe the inventories of organic and inorganic nitrogen to the soil. The organic inventory consisted of the existing organic nitrogen at the beginning of a month, plus nitrogenous waste added during the month, minus the nitrogen transformed to an inorganic form by net mineralization. Inorganic nitrogen was increasd by mineralization, plus additions of inorganic fertilizer, or inorganic nitrogen in the waste disposed on the land, minus loss due to leaching via ground water and consumption of nitrogen by growing plants. Potential losses to both inventories due to surface runoff, denitrification, volatilization and erosion, and possible gains from rainfall, decaying residues from plants and nitrogen-fixing organisms in the soil were all omitted in the model.

These omissions were assumed to be inconsequential relative to the monthly aggregates considered.

The principal transformation governing the inventories within the model was mineralization. This was described by the Van't Hoff-Arrhenius equation in which the rate of the reaction depends on temperature. It was necessary, therefore, to estimate average temperature in the soil. The transportation of nitrogen considered by the model was that due to leaching. This depends on water movement in the soil, and was predicted from estimated additions of water to the soil by rainfall, minus losses due to evapo-transpiration and leaching, again computed by the month. The equations were tested by comparing their predictions with observation.

In considering the disposal of wastes, the capacity of the crop to

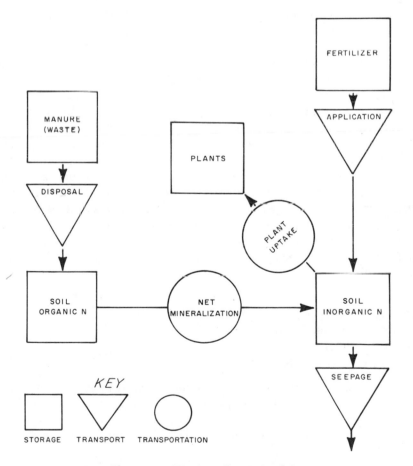

Figure 4.8. Nitrogen flow in Model A.

assimilate nutrients, the crop yields, must be taken into account. Both these factors are specific to the type of crop. Corn was selected for this model and regression equations were derived relating both the nitrogen consumed by corn, and yields, to the inorganic nitrogen in the soil.

The model was used to demonstrate the derivation of an optimum disposal schedule. This indicated how land disposal of nitrogenous wastes affects ground water quality, according to the assumptions of the model. The main conclusion was that crops are unlikely to assimilate all the nitrogen applied in the form of waste, and hence, as applications of waste increase, the concentrations of inorganic nitrogen correspondingly increases in the ground water.

General Assessment of Model A

Two levels of assessment which apply to all mathematical models can be distinguished. First, their purely mathematical structure and properties can be considered and second, the actual scientific assumptions and data can be assessed. Mathematical descriptions can be correct but results erroneous due to inadequate data, and the converse is also true.

Data and assumptions of Model A were confirmed by observations. The data requirements were moderate, making the model relatively accessible for general use. For the same reason it has heuristic value. However, two mathematical difficulties apply to this model. First, the equations governing mineralization, leaching and utilization of nutrients by crops were linearized. Hopefully, the use of monthly averaged data diminished any inaccuracies resulting from these approximations. Second, there was no sensitivity analysis, without which it is not easy to obtain direct insight into solutions by the linear program and how these vary with changing assumptions.

Model B: Nitrogen Control in Poultry Waste Management as Estimated by Simulation Modeling (Schulte 1974)

Description of Model B

The objective of this model was to quantify, by simulation, the nitrogen flow generated by a hypothetical poultry unit. In particular, the study assessed nitrogen losses to ground and surface waters following applications of poultry manure to land under various disposal policies.

As with Model A, the core of Model B was the nitrogen flow (Figure 4.9). Detailed equations simulated nitrogen budgets with gains and losses estimated by empirical equations taking into account environmental conditions—primarily soil moisture and temperature, rainfall, existing nutrient inventories and water movement.

The flow of nitrogen in the model originated with the conversion of protein in feed to the eggs and manure produced on the farm. Equations representing treatment of the manure were simple mass

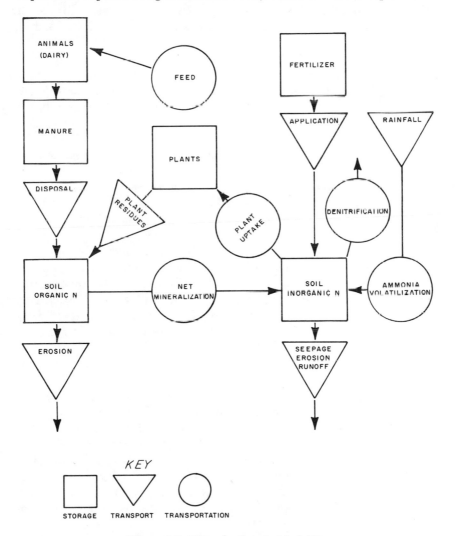

Figure 4.9. Nitrogen flow in Model B.

balances, which included mineralization and removal of nitrogen by denitrification and ammonia stripping. No attempt was made to simulate the microbiology of other transformations of nitrogen during treatment, such as urea hydrolysis and denitrification, because of incomplete knowledge of these processes. Further simplification was achieved by assuming that the manure was uniformly mixed in the plow layer of the soil, thereby eliminating ammonia volatilization. With data such as that presented in a previous section, this model could be readily expanded to account for losses due to ammonification from manure spread at different loading rates.

The soil nitrogen budget was represented by a comprehensive simulation of the transformations and transport due to mineralization, denitrification, leaching, crop uptake and runoff. Mineralization of organic to inorganic nitrogen was considered separately for nitrogen contained in the soil and in manure. The net mineralization of both was described by first order reactions with the Van't Hoff-Arrhenius equation being used to determine the rate, adjusted according to soil moisture.

Temperature and soil moisture also govern the first-order process by which denitrification in the soil is assumed to proceed. A rate is first estimated according to temperature, and then adjusted for soil moisture.

The cumulative uptake of nitrogen by plants is defined by a sigmoid curve, and in the model daily uptake was simply the difference between the total uptake on consecutive days. The model also allowed for the effect of drought on the utilization of nitrogen. Having estimated the amount of nitrogen utilized, expected crop yields could then be computed. To compute the losses of nitrogen due to runoff and leaching, the soluble inorganic nitrogen removed by surface runoff was calculated by multiplying the estimated concentration of nitrogen in rainfall by an enrichment ratio, and losses due to leaching were considered to be proportional to nitrogen in the soil, and to the volume of water flowing through it. Losses in eroded sediment were estimated from a modified form of the universal soil loss equation developed by Wischmeier and Smith (1965). This estimates soil loss as a function of rainfall intensity, soil erodibility, slope length, slope gradient, cropping management and erosion control. Having obtained an estimate of the movement of sediment, the corresponding loss of nitrogen was calculated by again using an enrichment ratio.

Additions to the soil inventories of organic and inorganic nitrogen from the residues of crop and rainfall, respectively, also were included in the model. Crop residues increased the magnitude of the soil organic inventory and hence became available for subsequent mineralization. Nitrogen added by rainfall was merely taken to be the volume of rain per unit area multiplied by concentration of nitrogen within it.

Both the transport and the rates of transformations from one form of nitrogen to another depend on the movement and availability of water. Model B included a detailed hydrological budget accounting for precipitation, runoff, evapotranspiration, seepage, and infiltration. Parallel to the nitrogen budget, equations representing the water balance and flow in the soil were applied. The accounting procedure used was a modification of the USDAHL–70 (Holtan, and Lopez, 1971) model of watershed hydrology.

Monte Carlo simulation was used to generate synthetic sequences of rainfall having the same statistical properties as the rainfall patterns of central New York State. The persistence within sequences of daily rainfall, i.e., the dependence between consecutive occurrences of rainfall, was described by a simple Markov chain. By use of the chain, the model determined the occurrence or non-occurrence of rain on any particular day. Given that rain was predicted to occur, then its magnitude was obtained by Monte Carlo simulation, using randomly generated numbers and monthly cumulative probability distributions of rainfall.

Evapotranspiration was also computed by means of a Monte Carlo simulation. Potential evapotranspiration on a particular day was calculated from the temperature simulated for that day, and then corrected for a day length factor.

Finally, the model included detailed equations depicting infiltration, percolation and runoff. Estimated infiltration in the water balance allowed for variations in storage due to surface irregularities in the land, and the infiltration capacity of the soil took account of foliage cover. The actual magnitude of subsequent percolation through the soil depends on the sum of evapotranspiration and infiltration and on the moisture content and moisture capacity of each soil level considered. If after the possible debits and credits, made according to these factors, the available water exceeded the capacity of the soil, then runoff was assumed to occur. The maximum storage possible in the soil had to be achieved in the simulation model before any runoff was permitted. Having simulated these various flows of water on and through the soil, estimates of the associated transport of nitrogen were calculated.

Assessment by Model B of Waste Management Policies

The objective of assessing waste management policies was to determine losses of nitrogen to ground and surface water following different management decisions. Three types of management decisions were considered: (1) the degree of nitrogen removed from poultry waste by treatment, (2) the rate of application to the land and (3) the timing of disposal. Considered in the simulations were 18 policies given by assumed rates of nitrogen removal of 25, 50, and 75% through treatment of the manure with application rates of 250, 500, and 1000 kg N/ha/yr, and disposal in the spring or daily. Each policy was subjected to identical simulated meteorological conditions.

The simulations demonstrated how losses of nitrogen to both surface and ground waters could be reduced if nitrogen were removed from the manure prior to disposal. For example, the estimated loss of inorganic nitrogen due to leaching was about 10 kg/ha for the policy representing 75% removal of nitrogen. This loss was nearly double the loss when nitrogen removal was assumed to be only 25%, for an otherwise similar policy. The results of the simulations also suggested that losses from daily applications of manure in the spring and fall were virtually indistinguishable from applications of the same total amount in the spring only, assuming high rates of nitrogen removal were imposed during treatment of the manure. However, applications in the autumn, of manure in which a low percentage of the nitrogen had been previously removed, substantially increased the inventory of inorganic nitrogen in the soil.

Similar conclusions were obtained with respect to runoff. For example, the total soluble inorganic nitrogen in runoff was negligible for all application policies if high levels of nitrogen were removed during treatment.

General Assessment of Model B

The computer simulation demonstrated the feasibility of constructing a detailed model of an agricultural unit of production. There are two major problems, however. First, large-scale simulations are notoriously tedious in their construction and use, requiring a large computing effort for their operation. Second, data requirements are correspondingly large, and the relative scarcity of reliable observations obtained in the field means that the determination of coefficients is difficult and sometimes arbitrary. It is important therefore to determine how sensitive the results of the model are to assumed coefficients. In the case of Model B, both the hydrological and nutrient submodels were comprehensively and successfully tested against field conditions. The most significant discrepancies between predicted and observed values were associated with major events, such as storms. However, difficulty in reproducing a single extreme event is to be expected in a large-scale mathematical model. When losses were aggregated over a period longer than a day, agreement between computed results and field observations was more satisfactory. Experience in mathematical modeling suggests that the smoothing process of aggregating, or lumping data and processes, more frequently provides this agreement. Model A is an example of the advantages in keeping simplification of structure and assumptions appropriate to the data and knowledge available.

Model C: Animal Waste and the Possible Impact of Legislation on Its Disposal. [Funded primarily under an EPA grant. Coote, 1973]

Description of Model C

A main aim of Model C was to assess hypothetical legislation for regulating the disposal of animal waste to control losses of nutrients. A complementary objective was to estimate the impact of legislative controls on farm income and the consequences for the quality of the environment.

There are two principal submodels in Model C. The first considers the nutrient flows of both nitrogen and phosphorus, the water flow, and relevant economic factors.

The second part is an application of a linear programming procedure superimposed, on the first submodel, to compute the best way of meeting the given objectives. The unit of time used by the model was three months.

Nutrient and water flows: The unit of production investigated by Model C was the dairy farm. Initially the model represented feed requirements for the cattle in terms of total digestible nutrients and protein. Resulting waste was expressed as a function of the number, type and age of animals and the period of time considered. As the model was developed, provision was made for storage but not treatment of the waste. As with preceding models, the nutrient flow in the soil was represented by assumed nutrient inventories. Sources of nutrients were fertilizer and manure. Nutrient requirements essential for crop production were estimated and fertilizer applications computed by taking the difference between the minimum expected nutrient content of manure and the total requirement of the crop. Rainfall was also included as a source of nitrogen. (Figure 4.10 shows the assumed flow of nitrogen.)

Losses of nitrogen and phosphorus representing potential pollution were summarized by two equations. These provided for estimated nitrogen losses due to ammonia volatilization, denitrification, runoff, percolation and soil erosion. Soil erosion was considered to be the only means by which phosphorus was lost from the soil. Such losses represent total phosphorus and, as suggested earlier in this report, estimates of losses of total phosphorus may be of limited

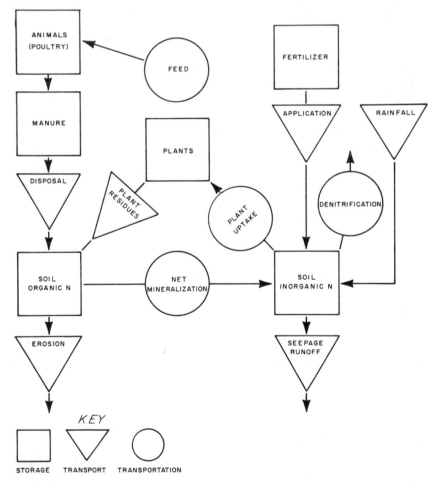

Figure 4.10. Nitrogen flow in Model C.

use since biologically available phosphorus, a fraction of the total, is of primary concern for management.

As with Model B, the gains and losses of nutrients representing the nutrient flow were superimposed on corresponding hydrological factors, including rainfall and runoff. Losses due to percolation and denitrification were taken to be the difference between a computed total potential loss and actually computed losses. The runoff model, basically that of the Soil Conservation Service (Mockus, 1971), was a function of rainfall storms, crop, drainage class of the soil, and time. Since this did not include land gradient as a variable, the model was modified to allow for an increase in runoff as the slope increased. Seasonal losses of nitrogen and phosphorus due to sediment movement were estimated from the Universal Soil Loss Equation.

Economic factors: Detailed linear equations described various economic inputs to the farming unit, and corresponding farm income. Included were labor and production costs and the expected return from marketing crop and animal products. Costs and revenue were subdivided according to the main farming activities of animal raising, waste handling, soil management and crop production.

Optimization model: Factors developed above were subject to various constraints in the optimization model. Briefly, these constraints specified the maximum amount of land that may be utilized for farming operations, labor input with the condition that seasonal variations in the hiring of labor were excluded, and size of the herds, with the quantity of feed they consumed. The model was also constrained so that solutions were consistent with existing production practices and crop rotations.

Concerning conditions governing nutrient losses, or pollution control, two main types of legislated constraints were considered. First, the levels of manure disposed on the land had to be within specified limits. Second, controls were achieved through zoning for land use.

The effect of assuming hypothetical legislation requiring controls, as indicated above, was assessed through applications of the model. The results indicated that a major consequence of such legislation was to limit the acreage of land available for waste disposal. At herd densities of 1 cow to 4 hectares or less, combined controls for manure and fertilizer applications were able to limit the losses of nutrients to surface and ground water without affecting farm income. If the ratio of cows to unit area of land were increased, then total losses of nutrients per unit area increased and farm income were reduced. Higher densities of animals require more land for feed production, incurring an increased possibility of sediment, and hence nutrient, loss. Also the volume of waste to be disposed increases relative to the land used for animal production as density increases.

The results also indicated that under severe restrictions, some land could perforce be retired from agricultural use to be used solely for waste disposal. This necessarily reduces cow/land ratios to approximately 0.3 cows/ha. In some cases, a change in cropping practices would suffice to meet the assumed controls. For example, the substitution of grass for alfalfa increases the area of land available for disposal. Or, increasing the acreage used for small grains at the expense of corn, extends the period of time within a year during which manure can be spread. However, such changes may reduce farm income.

General Assessment of Model C

A principal objective of this model was to evaluate the consequences for both farm incomes and the environment, in terms of nitrogen and phosphorus losses, if farm incomes were maximized within the specified controls. Although the model used the specific example of dairy farming and the disposal of cow manure, its procedures are generally applicable.

The scientific core of the model accounted for nutrient and water flow. Some effort was made to substantiate the data and assumptions describing these natural phenomena. As previously stated, the lack of field data was a handicap, but provisional validation for the predictions for runoff seepage and soil losses was achieved. Data obtained from the water quality plots at the Cornell's Aurora Research Farm were used, and although discrepancies were discovered, the predictions of the model and the field observations were in general agreement. A major discrepancy was in the predictions for soil loss. Unfortunately the field data had been obtained over a period including very high rainfall and the form of the soil loss equation used could not completely account for very wide deviations from average conditions.

Model D: Economic Analysis of Policies to Control Nutrient and Soil Losses (Schaffer, 1974)

Description of Model D

This model was designed to assess the economic consequences of reducing losses of nitrogen and phosphorus from agricultural activities in watersheds. Specific objectives were to relate agricultural production to nutrient and soil losses affecting water quality, and to consider and compare the economic consequences of controls. In some respects, this model represents an extension of the investigations already discussed. As in Model C, there are two main submodels: (1) a highly detailed simulation of environmental factors governing the daily nutrient and water flows through the catchment; and (2) an economic model which applied a linear programming model to aggregated data representing annual values.

Accounting procedures used in the environmental submodel were the same in principle as those in Model B. The submodel had itself four submodels delineating the flow inventories: soil water, soil erosion, and nitrogen and phosphorus flows. The last two were in part superimposed on the second, and all three were superimposed on the first. As in Model C, principal factors governing the soil water inventory were daily air and soil temperature, rainfall, evapotranspiration, runoff, infiltration and seepage. The determination of the balance of these last factors allowed the estimation of runoff, which was computed directly from the difference between rainfall and infiltration.

A modified form of the universal soil loss equation suggested by Meyer and Wischmeier (1969) provided the basis of the submodel representing soil erosion. The original equation was developed to predict annual losses of soil from relatively small land area. Since the simulation of Model D represented daily events, it was believed that a better representation of total soil erosion would be obtained if cropping factors, which clearly change within a year, were allowed for. A further adjustment allowing for differences in soil erodibility was also made.

The double entry bookkeeping system summarizing the gains and losses of nitrogen in the nitrogen submodel may also be considered to be a refinement of the accounting procedure in Model B (Figure 4.11). Mineralization processes for urea hydrolysis, ammonification, immobilization and nitrification were all specifically included together with volatilization. In particular, detailed consideration was given to mineralization and immobilization, and the dependence of the latter on the carbon-nitrogen ratio.

Urea hydrolysis was assumed to be identical for urea in both manure and in fertilizer, and was described by a regression equation relating hydrolysis to time and temperature of the surface soil. When ammonia is produced it is easily volatilized as shown earlier in this chapter, and another regression equation expressed this loss as a function of time, temperature, superphosphate, and straw if

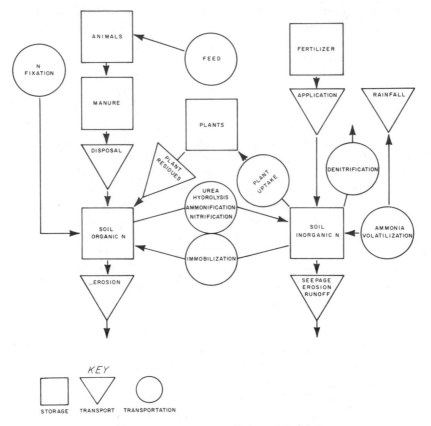

Figure 4.11. Nitrogen flows in Model D.

present in the soil. Similar regression equations described the immobilization of nitrogen as a function of soil pH, soil temperature, and soil moisture, and nitrification. Nitrification itself depends on these variables, and in addition, on the amount of ammonia present in the soil.

A much less detailed submodel was used to describe the flow of phosphorus and no transformations and transfers corresponding to the nitrogen submodel were defined. The loss of phosphorus was related to the amount of manure present on the surface of the soil and soil loss due to runoff.

The environmental submodel was operated assuming a distribution of hourly rainfall, together with a distribution representing short-term intensities. It was then possible to produce coefficients which estimated losses of nitrogen, phosphorus and soil for various crops and types of soil. Having obtained these coefficients from the environmental submodel, they were incorporated in a linear programming model which included the economic factors.

This overall model was applied to a small watershed in the drainage basin of Canadarago Lake. Considerable simplification of the data was necessary. For example, the number of soil types was reduced to six and to each type was ascribed one level of crop productivity. Given these assumptions for soil parameters in the watershed, and the appropriate costs and revenue, the impact of three main types of management policies were postulated. These were: constraints limiting the losses of nutrients, hypothetical taxes on non-point or diffuse wastewater, and taxes on fertilizer.

The major conclusions were as follows. Farm income falls when controls on nutrient losses are imposed. The more the losses have to be reduced, the greater is the fall in net farm revenue. Regulations applied to either nitrogen or phosphorus results in reduction in losses of the other nutrient and soil because the losses of nutrients and soil are interdependent. A reduction of the amount of phosphatic fertilizer applied within a watershed does not necessarily produce a reduction in phosphorus in drainage water. It follows, therefore, that a fertilizer tax on phosphorus may not be an effective means of limiting water pollution.

Assessment of Model D

Of the four models, Model D was the most ambitious. It may, therefore, be argued that this model more seriously incurs the risk of departing from the available data and knowledge. In an effort to obtain usable mathematical relations, many regression equations were calculated using data from investigations that may not have been directly relevant to the model. However, given the lack of alternatives, this was perhaps unavoidable. Because the loss coefficients were based on equations designed to estimate losses from small plots, the estimated nitrogen and phosphorus losses exceeded the levels of nutrients measured in the stream. Hence the ratio of observed to calculated losses, i.e., the delivery ratio, would normally be less than one. Ratios of 0.3 and 0.8 for nitrogen and phosphorus, respectively, were used to adjust the calculated losses. None of the other models reviewed attempted to make such adjustment.

With regard to the management policies considered by the model, it is doubtful that a tax on non-point wastewater is practicable. There are difficulties in adequately sampling well-defined single point sources of pollution, and the position regarding non-point sources is even more complex.

Concerning the mathematics of the model, the form of some of the equations used to portray processes in the environmental sub-model were not entirely satisfactory as was possibly true of all the models. In some cases, there were multicollinearities between the independent variables with the possible consequence that the estimated coefficients were misleading. Also, the environmental model did not attempt to explicitly represent the hydrology of a watershed. Clearly, the flow of nutrients and sediment is dependent on the complete hydrological flow and a model which directly takes such flow into account might produce different results. Allied to this question is the fact that Model D basically constructed a watershed model as an extension of the field plot. Given existing knowledge, however, this may be the most appropriate and possibly the only method.

Despite reservations, the model as a whole is a useful instrument for producing from existing data an evaluation of management policies that would otherwise be very difficult to quantify. Both the mathematical structure and data input can be readily modified as knowledge advances. Equally, other management policies can be formulated and assessed.

CONCLUSION

All the models confirm common sense in that their results showed that nutrient losses from agriculture cannot be entirely eliminated. However, some practical steps may be taken to reduce the rate of losses. Man's intervention in terrestrial flows of nitrogen and phosphorus has served to augment the quantities transferred from land to air and water. It is therefore highly doubtful that without eliminating human activity entirely from an area, the level of losses can be restored to those prevailing before the activity commenced. To some extent the models discussed indicated that the most effective way to reduce losses from farm land is simply to retire that land from farming. It is unlikely that this solution will be universally accepted.

A more reasonable objective would be to encourage methods of cultivation and nutrient application which reduce losses with least financial penalty and possibly with a gain, for the farmer, such as indicated in the Long Island study. Similar objectives can also be applied to the treatment and storage of manure prior to disposal as outlined in the following chapters.

REFERENCES

Coote, D. R., D. A. Haith and P. J. Zwerman. 1975. The environmental and economic impact of nutrient management on the New York dairy farm. Search (forthcoming).

Haith, D. A. 1973. Optimal control of nitrogen losses from land disposal areas. J. Environ. Engng Div. EE6. pp. 923–935.

Holtan, H. N. and N. C. Lopez. 1971. USDAHL–70 model of watershed hydrology. Technical Bulletin No. 1435. USDA Agricultural Research Service.

Lathwell, D. J., D. R. Bouldin and W. S. Reid. 1970. Effects of nitrogen fertilizer application in agriculture. In Cornell University Waste Management Conference, Rochester, N.Y.

Likens, G. E., F. H. Bormann, N. M. Johnson, D. W. Fisher, and R. S. Pierce. 1970. Effects of forest cutting and herbicide treatment on nutrient budgets in the Hubbard Brook watershed-ecosystem. Ecol. Monogr. 40:23–47.

Likens, G. E. 1972. The chemistry of precipitation in the Finger Lakes Region. Techn. Rep. No. 50. Cornell University Water Resources and Marine Science Center, Ithaca, N.Y.

Maynard, L. A. and J. K. Loosli. 1969. Animal Nutrition. McGraw-Hill Book Co., Inc. 6th Ed.

Meyer, L. A. and W. H. Wischmeier. 1969. Mathematical simulation of the process of soil erosion. Trans. Am. Soc. Agr. Engn.

Mockus, V. 1971. Hydrology. National Engineering Handbook, Sect. 4. S.C.S., USDA, Rev.

Sawyer, C. N. 1965. Problems of phosphorus in water supplies. Jour. Amer. Wat. Works Assoc. 65(11): 1431.

Schaffer, W. H., J. J. Jacobs and C. L. Casler. 1974. An economic analysis of policies to control nutrient and soil losses from a small watershed in New York State. In Processing and Management of Agricultural Wastes. Cornell University, Ithaca, New York.

Schulte, D. D., R. C. Loehr, D. A. Haith and D. R. Bouldin. 1974. Effectiveness of nitrogen control in poultry waste management as estimated by simulation modeling. In Processing and Management of Agricultural Wastes. Cornell University, Ithaca, New York.

Stevenson, C. K. and C. S. Baldwin. 1969. The effect of time and method of nitrogen application and source of nitrogen on the yield and nitrogen content of corn. Agron. 61:381–386.

Vallentyne, J. R. 1974. The algal bowl: Lakes and man. Department of the Environment, Fisheries and Marine Service. Ottawa, Canada.

Vitousek, P. M., and W. A. Reiners. 1975. Ecosystem succession and nutrient retention: A hypothesis. BioSci. 25(6):376–381.

Wischmeier, W. H. and D. D. Smith. 1965. Predicting rainfall-erosion losses from cropland east of the Rocky Mountains. U.S. Department of Agriculture. Handbook 282. Washington, D.C.

ACKNOWLEDGMENTS

Dr. D. A. Haith, Dr. W. J. Jewell and Dr. C. A. Shoemaker all reviewed the section on the Cornell models and made helpful suggestions.

Principal Authors

K. S. Porter, D. A. Lauer, J. Messinger, D. R. Bouldin.

5

Economic Analysis of Reducing Phosphorus Losses from Agricultural Production

Chapter 5

ECONOMIC ANALYSIS OF REDUCING PHOSPHORUS LOSSES FROM AGRICULTURAL PRODUCTION

The limit on improving water quality in our regions is essentially an economic one. Although further technological advance may allow better quality with a smaller expenditure of resources, the controlling question will continue to be, 'How much of society's resources shall we devote to maintaining and improving water quality?'

—Allen V. Kneese and Blair T. Bower

INTRODUCTION

The research reported in Chapter 2 indicates that the production of phytoplankton in lakes such as Cayuga is closely associated with loading of biologically available phosphorus. Only a relatively small fraction of the phosphorus leaving Fall Creek is biologically available and of major significance to aquatic biology. Data in Chapter 3 demonstrate that much of the phosphorus entering Cayuga Lake from Fall Creek is particulate phosphorus carried on sediment. Models that predict inputs of total phosphorus to lakes are likely to be misleading in terms of effects on productivity because the proportion of phosphorus that is biologically available varies among watersheds. In this chapter an attempt is made to consider biologically available rather than total phosphorus.

Data presented in Chapter 3 indicate that nitrogen levels in Fall Creek are well below the United States Public Health Service standard at all times. Therefore, little emphasis was placed on the cost of controlling nitrogen losses to Fall Creek and Cayuga Lake.

The purpose of the work described in this chapter was to estimate the costs of reducing phosphorus inputs to Cayuga Lake from the Fall Creek watershed. No attempt was made to estimate the benefits of reducing phosphorus inputs to the lake. It is recognized that

169

such a reduction may produce both positive and negative benefits to users of the lake. From the standpoint of economic efficiency, the quality of water in Cayuga Lake should be improved as long as the additional benefits from improved water quality exceed the additional costs. Benefits from reduced phosphorus inputs may occur from less algal production, which makes the lake more enjoyable for boating and swimming and more suitable as a supply of potable water. On the other hand, a reduction in phosphorus input may reduce fish production, thereby reducing benefits to fishermen. While a clear understanding of benefits resulting from a reduction of phosphorus inputs awaits further study, costs associated with phosphorus reduction from various sources can be estimated.

OBJECTIVE OF THE ECONOMIC ANALYSIS

The major objective was to estimate and compare the costs of reducing phosphorus losses to Cayuga Lake from agricultural production and municipal sewage in the Fall Creek watershed. The emphasis was on determination of the costs of reducing phosphorus losses from farming. In estimating these costs, three sources of phosphorus were considered: (1) land runoff as related to soil erosion, (2) land runoff as related to manure applications, and (3) barnyard runoff. In addition, possible trade-offs between nutrient losses to water and other environmental characteristics were considered.

LAND RUNOFF AS RELATED TO SOIL EROSION

The cost of reducing phosphorus losses from land runoff was estimated by using a linear programming model of agriculture in the Fall Creek watershed. The development of such a model requires information on (1) production alternatives, (2) costs and returns associated with each alternative, and (3) the phosphorus loss from each production practice. The data sources used in developing the model, the procedure developed to estimate phosphorus losses, formulation of the model, results, and conclusions are discussed in subsequent sections.

Data Sources and Model Development

Nearly all farms in the watershed are dairy farms. The locations of these dairy farms were plotted on topographic maps. The crop grown on each field in the watershed was identified while driving all roads in the area and recorded on air photos. Of the 130 dairy farms in the watershed, 20% were surveyed in 1973 to collect information on cow and heifer numbers, milk production, crop acreages and yields, and fertilization and manure handling practices. From the survey it was estimated that there were 6800 cows in the watershed.

The boundaries of each farm surveyed were drawn on air photos. By overlaying the air photos with the appropriate soil maps, the acreages of crops by soil types and slopes were tabulated. Over 130 combinations of soil types and slopes were found. To reduce these to a tractable number, they were consolidated into 19 groups, each of which was believed to have similar characteristics and yield levels. These soil groups and crop acreages were proportionately expanded to cover 100% of the farming in the Fall Creek watershed (Figure 3.1, p. 64). The small amount of farming below Freeville was not considered in the model.

The crops considered in the watershed model were corn silage, corn grain, oats, predominately legume hay, predominately grass hay, and pasture. Small amounts of other crops found in the survey were disregarded. Variable costs and labor requirements were estimated for growing and harvesting each crop on each soil type. Rates of N, P_2O_5, and K_2O fertilization and manure application were established for each crop, based on averages derived from the survey (Table 5.1). Yield levels [Table 5.2 and 5.3(a)] for each crop on

Table 5.1. Variable costs, labor requirements and fertilizer requirements per hectare of crops per year.

Item	Unit	Corn Silage	Corn Grain	Oats- Reseeding	Legume- Grass	Grass- Legume
Variable costs	$	79	74	215[1]	52	32
Labor	hr	19.8	14.8	14.8	19.8	14.8
Nitrogen	kg	84	90	11	0	6
P_2O_5	kg	67	62	34	22	0
K_2O	kg	62	56	34	34	0
Manure	mt	36	20	—	—	7

[1] Includes lime requirement for the rotation. In the case of continuous corn, a lime cost was charged to corn.

each soil type were established from the Soil Survey Reports for Tompkins, Cortland, and Cayuga counties (Neeley, 1965; Seay, 1961; Hutton, Jr., 1971, respectively). The crops were combined into nine rotations. The intensity of rotations, in terms of percent-

Table 5.2. Yields in metric tons per hectare in linear programming model.[1]

Groups	Yield Level	Silage[2]	Grain	Oats	Legume Hay	Grass Hay	Permanent Pasture
18	1	38.1	5.9	2.6	9.5	6.2	3.1
1,2,4,5	2	35.9	5.5	2.5	9.0	5.6	2.8
3,6,7,8,	3	31.4	4.8	2.2	7.8	4.5	2.2
9,11,12,14,15	4	26.9	4.1	1.8	6.7	3.4	1.7
10,13,16,17,19	5	22.4	3.5	1.5	5.6	2.2	1.1

[1] Corn silage, corn grain and oat yields were reduced by 10% and hay yields by 20% to cover harvesting, storage and feeding losses.

[2] Corn silage contains 25% dry matter; the other crop yields are computed on the basis of 85 to 90% dry matter.

age of corn, for each soil type and slope group was limited in the initial solution to that found on the farms in the survey. For each combination of crop rotation, soil type and slope, the losses of soil, nitrogen, and phosphorus to water were estimated.

Estimation of Soil and Phosphorus Losses

No generally applicable nutrient loss coefficients for specific soil types and production practices were available. An attempt was made, using regression analysis of plot data from several locations, to estimate the relationship between nitrogen and phosphorus losses and runoff and/or soil erosion. Satisfactory regression equations were found for some situations but it was not possible to make reasonable estimates of nutrient losses for the Fall Creek watershed. A major reason for this was that values of the independent variables, soil loss or runoff, for some soils in the watershed fell outside the range of the experimental data used in estimating the regression equations. After considerable work, this effort was abandoned and the decision was made to use the universal soil loss equation (Wischmeier and Smith, 1965) and nutrient enrichment ratios.

To develop plans for controlling soil loss, knowledge of the physical relationship between manageable factors in the production process and soil loss is required. Toward this end, the "universal soil loss equation" provides a procedure for computing the expected average annual soil loss from alternative practices on a particular land area. The application of this equation gives estimated long-term average annual soil losses (25 years or more)

Table 5.3(a). Dry matter and nutrients produced in kg per hectare of crops, yield level 3.[1]

Item	Corn Silage	Corn Grain	Oats	Legume Hay	Grass Hay	Permanent Pasture
Dry matter	8473	4489	1705	5649	3228	1614
Total digestible nutrients	5931	3856	1356	3389	1793	897
Protein	712	412	223	847	305	152

[1] Yield level 3 is slightly above the average yield levels found in the watershed survey. Dry matter and nutrient production of other yield levels shown in Table 5.2 are proportionately above and below production shown for level 3.

Table 5.3(b). Numbers of cows and replacements and annual feed requirements, Fall Creek watershed.

	Number	Total Digestible Nutrients	Total Protein	Dry Matter
		kg	kg	kg
Dairy cows	6800	3810	816	5543
Replacements	2040	3010	634	5942

caused by erosion of soil by water. In predicting losses from a particular land area, the equation takes into consideration rainfall intensity and duration, soil type, slope length and gradient, cropping practice and erosion-control practices. These factors make up the universal soil loss equation (Wischmeier and Smith, 1965):

$$A = RKLSCP$$

where,

A is the computed soil loss in tons per acre per year,
R the rainfall factor,
K the soil-erodibility factor,
L the slope-length factor,
S the slope-gradient factor,
C the cropping-management factor, and
P the erosion-control practice factor.

The R factor is a measure of the erosive force of a normal year's rainfall. The R value for the area including the Fall Creek watershed is 100 (Swader, 1974). The K factor for a given soil type is determined under a cultivated, continuous fallow condition. The K factor values for the various soil types were obtained from Swader (1974).

Both slope length and gradient are important in determining soil erosion. LS values were computed by solving the following equation:

$$LS = f^{\frac{1}{2}} (0.0076 + 0.0053s + 0.00076s^2)$$

where f is the slope length in feet and s is the gradient in percent. Slope length and gradient values for the soil types considered in the study were obtained from Hansen (1974).

RKLS estimates the average annual soil loss for an area under a continuous fallow condition. This would represent maximum erosion, which can be reduced through crop, tillage, or erosion control practices.

C factors for the nine cropping practices considered were derived from Wischmeier and Smith (1965). In addition to the nine cropping practices, contour strip cropping was considered on slopes where this practice was considered to be effective. Values for contour strip cropping were obtained from Wischmeier and Smith (1965).

The next step was to develop a procedure for estimating erosion losses of phosphorus. Three factors were given prime consideration (Jacobs, 1972):

1. The positive relationship between erosion and phosphorus losses,
2. The relative immobility of phosphorus in soil, and
3. The selective removal of phosphorus by erosion.

Taking these characteristics into consideration, the phosphorus loss equation used was:

$$P = A_1 TE$$

where,
P is the kg of phosporus loss per hectare per year,
A_1 is the mt of soil loss per hectare per year (A converted from tons/acre to mt/ha),
T is the kg of phosphorus per mt of topsoil, and
E is the enrichment ratio, *i.e.*, the increased concentration of phosphorus in the eroded soil relative to the original topsoil.

To apply this concept of linking soil and phosphorus losses, two additional pieces of information are needed. First, the total phosphorus content of the eroded soil is needed as a multiplier times the quantity of soil loss to obtain phosphorus loss. A value for T of 0.39

kg per ha of topsoil was calculated from Buckman and Brady (1960). Second, soil erosion processes selectively remove smaller lower density soil particles. These particles have a greater reactivity than other fractions of soil solids and therefore react with phosphorus more readily. The result is a higher content of phosphorus in the eroded soil fraction compared to the bulk soil. The ratio of phosphorus content in the eroded soil to the phosphorus content of the original soil is denoted the "enrichment ratio." Furthermore, enrichment ratios are inversely related to the quantity of soil loss as shown in Figure 5.1. This enrichment ratio curve was developed

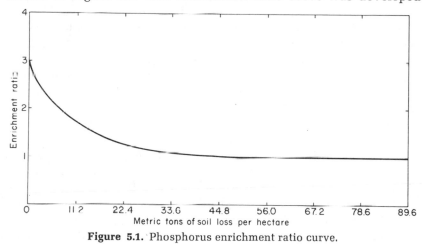

Figure 5.1. Phosphorus enrichment ratio curve.

from Jacobs (1972). The appropriate enrichment ratio is then multiplied times the product of soil loss and bulk soil phosphorus content to obtain total phosphorus loss from soil via erosion.

The above procedure was used to estimate phosphorus loss for each combination of production practice and soil group considered. This procedure was also used to estimate nitrogen losses associated with surface runoff. However, because nitrogen levels in Fall Creek water are well below USPHS standards, no restrictions were placed on nitrogen losses.

Application of Soil and Nutrient Loss Equations to the Fall Creek Watershed

Soil and phosphorus losses were computed for the combinations of crop rotations and conservation practices allowed on each of the 19 soil groups. The computed losses from alternative production practices are a weighted average for the soil types in that group.

A disadvantage of the soil and nutrient loss equations is that they estimate the gross quantity of soil and nutrients moved from their original position. Since the primary concern is with only that portion of soil and nutrients entering a water course, the initial soil and nutrient loss estimates must be adjusted for the soil redeposited within the landscape of the watershed during overland flow. At present, there are no functional relationships developed to describe the deposition of eroded soil during overland transport. A factor commonly used to correct empirically for the efficiency of sediment transport from a watershed is the delivery ratio. The delivery ratio is the quotient of measured sediment deposits in a lake receiving water from a watershed and the soil erosion predicted for that watershed by a soil loss equation.

In attempting to determine soil delivery ratios, some studies have tried to correlate delivery ratios with drainage area. Generally, an inverse relationship is indicated. Data from widely scattered areas show that soil delivery ratios throughout the country roughly vary inversely as the 0.2 power of the drainage area (United States Environmental Protection Agency, 1973). Using this relationship, a delivery ratio of 0.1 was obtained for the Fall Creek watershed. A delivery ratio of 0.2 was also used to check the sensitivity of the model results to the delivery ratio.

The combination of the universal soil loss equation, nutrient enrichment and delivery ratios discussed above was used to estimate phosphorus losses from the watershed. This estimated phosphorus is largely, but not entirely, particulate phosphorus. It is generally thought that only a small fraction, probably 5 to 10%, of the phosphorus loss related to soil erosion is biologically available (Keup, 1968; Stanford, et al., 1970). The significance of controlling biologically available phosphorus rather than total phosphorus is discussed in later sections.

Linear Programming Model

The linear programming model used to estimate the cost of reducing nutrient losses from land runoff as related to soil erosion includes a set of production activities and restrictions for the watershed. Production activities considered were dairy cows, raised and purchased replacements, alternative crop rotations for each soil type, purchased grain and protein supplement, purchased fertilizer and hired labor. The restrictions on the model include the hectares of each soil group in the watershed. Hectares of corn grain and corn silage grown and the hours of family labor available were restricted to the amounts in the watershed in 1973. Additional equations ensure that the requirements of the dairy cows and replacements for total protein and total digestible nutrients are met with either home-grown or purchased feed.

The model was restricted so that the initial solution closely depicted the crop and livestock production activities (Tables 5.3(b) and 5.4) and net farm income in the Fall Creek watershed in 1973.

Table 5.4. Farmland by soil group in Fall Creek watershed (1973) based on a survey of 20% of the farms in the watershed.

Group	Major Types	Slopes	Total Hectares	Crop Hectares	Pasture Hectares	Corn Hectares
1	Valois-Howard Langford-Howard	B	511	329	103	135
2	Bath-Valois	C	2,854	1,926	391	729
3	Bath-Valois	D	426	229	105	38
4	Howard	A	40	25	11	11
5	Howard Howard-Valois	C	943	652	163	267
6	Howard Valois-Howard	D	328	117	124	47
7	Langford, Mardin	B	2,493	1,842	302	494
8	Langford, Mardin	C	1,507	761	398	247
9	Langford, Mardin	D	632	243	121	30
10	Lordstown, Bath	E	524	114	104	17
11	Lordstown	B	805	148	26	55
12	Lordstown	C	309	148	65	55
13	Lordstown	D	261	102	70	7
14	Volusia, Erie	B	1,901	951	408	375
15	Volusia, Erie	C	765	237	321	68
16	Volusia	D	81	2	38	0
17	Wayland, Sloan Alluvial land	A	781	177	263	76
18	Genesee, Eel	A	306	197	47	104
19	Chippewa, Ellery	A	244	37	205	5
	Total		15,711	8,237	3,265	2,760

Comparison of Estimated Soil and Phosphorus Losses with Fall Creek Stream Data

In the initial solution, estimated soil loss from crop production in the Fall Creek watershed was 74,076 mt. A sediment delivery ratio of 0.1 results in an estimate of 7408 mt of sediment leaving the watershed in an average year which is about 50% of the 14,000 mt of sediment estimated in Chapter 3. It was also estimated in Chapter 3 that 40 to 60% of the sediment is due to human activity, including agriculture, construction, and other activities. Therefore, an estimate of 7408 mt seems high for agriculture alone, suggesting that a delivery ratio of 0.1 is too high or that the soil loss equation overestimates soil erosion in the watershed.

Estimated phosphorus loss associated with the 74,076 mt of sediment from crop production was 56,700 kg per year. Only part of this leaves the watershed. The delivery ratio of phosphorus is unknown and can only be estimated.

Because BAP rather than phosphorus is the important factor in aquatic production, the proportion of the estimated phosphorus loss from farming that is biologically active becomes important. The estimated annual loss of BAP from farming in Fall Creek for three delivery ratios and three percentages of BAP in relation to phosphorus are shown in Table 5.5. For comparison purposes, the

Table 5.5. Estimates of phosphorus and "biologically available phosphorus" related to soil erosion losses from Fall Creek watershed for several delivery ratios and solubilities.

Delivery Ratio	kg Phosphorus	kg "Biologically Available Phosphorus" at Various Percentages BAP		
		5	10	20
0.1	5,670	283.5	567	1,134
0.2	11,340	567	1.134	2,268
0.4	22,680	1,134	2,268	4,536

estimated annual losses of TSP from the watershed, based on Chapter 3, are shown in Table 5.6. The estimate of TSP from all diffuse sources suggests that losses related to soil erosion from cropland in each of the two years were less than the 600 and 1600 kg shown. BAP would be approximately 10% greater than TSP. Considering the fact that there must be BAP losses from manured fields and manure in barnyards, a 0.1 delivery ratio and 20% BAP or a 0.2 delivery ratio and 10% BAP would be reasonable in relation to the results presented in Chapter 3.

Results

The initial solution of the watershed linear programming model was computed with no restrictions on nutrient losses. In subsequent

Table 5.6. Estimated sources and loading of total soluble phosphorus for two one-year periods, Fall Creek watershed.

Source	September 1972–August 1973	May 1973–April 1974
	kg	kg
Biogeochemical (nonhuman)	3,100	2,320
Human		
Point	1,840	1,190
Diffuse	1,600	600
Total	6,540	4,110

Source: Table 3.16

solutions, losses of phosphorus were restricted by increments of 10% of the phosphorus loss predicted by the initial solution. The model then computed the least-cost rearrangement of the production activities to achieve each phosphorus restriction. In each case, the reduction in phosphorus loss continued until further reduction was not possible. The reduction in net farm income due to each phosphorus restriction is the cost to farmers in the watershed of decreasing phosphorus inputs to Cayuga Lake.

Phosphorus restrictions were placed on the model under four sets of conditions. These are referred to as Models I through IV.

Model I. All land currently in crop production must continue to be cropped.

Model II. Cropland is allowed to be idle, but idle land has soil and nutrient losses.

Model III. Conservation practices (contour strip-cropping) are introduced on certain soil types and slopes. In addition, land is allowed to be idle as in Model II.

Model IV. In addition to the conservation practices and idle land allowed in Model III, purchase of hay, grain, and replacements from outside the watershed are restricted to the levels in the initial solution.

The results are described below. In each case, solutions for phosphorus delivery ratios of 0.1 and 0.2 were computed. In all cases, both annual phosphorus losses and annual costs to farmers are estimated, rather than actually measured. Limitations of the analysis are discussed following presentation of the results.

Model I

In Model I, all land in the watershed cropped in 1973 was required to be cropped under the phosphorus restrictions. As the phosphorus loss was restricted by increments of 10% of the initial 56,700 kg (11,340 and 5,760 kg for delivery ratios of 0.2 and 0.1, respectively), net farm income in the watershed was reduced. This reduction represents the cost to farmers of reducing phosphorus losses (Figure 5.2 and Table 5.7). The cost for a given percentage

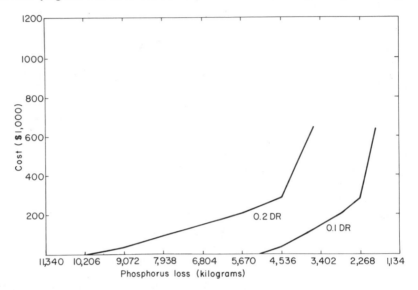

Figure 5.2. Cost of reducing phosphorus loss from farming in the Fall Creek watershed—Model I.

reduction is the same regardless of the delivery ratio (DR). The differences between alternate delivery ratios are the actual amount of reduction in phosphorus loss and the amount of phosphorus loss remaining after a given percentage reduction. The reduction of phosphorus losses is achieved by a reduction of corn acreage and an equivalent increase in hay acreage. Less total digestible nutrients are produced by hay than by corn; therefore, additional hay or grain is purchased to supply the required nutrients. The number of cows and raised replacements in the watershed did not change. The cost to farmers of reducing phosphorus is equivalent to the cost of buying additional feed, less the reduction in production costs if hay rather than corn is grown.

Model results indicate that the first 10% reduction in phosphorus loss could be achieved at no cost to farmers. In fact, there would be a small increase in income. This occurs because the amount of corn grown on steep slopes was decreased and the amount grown on

Table 5.7. Estimated cost of reducing phosphorus loss from farming in Fall Creek watershed, Model I.

% Reduction in P	% Reduction in Corn	Cost to Watershed	Cost per Farm	% of Net Income
10	0	$ 0	$ 0	0
20	6	35,700	275	3
30	17	88,200	678	7
40	30	148,000	1,138	11
50	48	206,000	1,585	16
60	72	285,900	2,199	22
68	100	640,500	4,927	49

more gentle slopes increased from the amounts found in the watershed. In reality, this would be difficult to achieve because the steep and more gently sloping soils are not necessarily on the same farms. Considering this situation, it was decided to assume that the cost of achieving the first 10% reduction in phosphorus was zero.

After the first 10% reduction, the cost of reducing phosphorus loss increased at an increasing rate as the phosphorus restriction was raised by increments of 10%. The cost of the second 10% reduction was $35,700 for the watershed or $275 per farm, approximately 3% of the estimated net income* per farm in 1973 (Table 5.7). A 50% reduction would cost farmers in the watershed $206,000 or $1585 per farm, 16% of net income. The cost of the largest possible reduction in phosphorus, 68%, was 49% of net income per farm.

With a 0.2 delivery ratio, the cost of each 10% reduction in phosphorus is the same as the cost with a 0.1 delivery ratio. However, the reduction in phosphorus loss is twice as large. The lowest achievable loss is also twice as large with a 0.2 delivery ratio as with a 0.1 delivery ratio.

Model II

In Model II, cropland was allowed to be left idle. Losses from idle land were estimated to be equal to losses from permanent pasture on the same soil type. The cost to farmers for a given reduction in phosphorus loss was slightly less than in Model I (Tables 5.7 and 5.8 and Figures 5.2 and 5.3). It was cheaper to leave some land idle and buy feed rather than to operate the land. In addition, the

---

* Net income was defined as labor and management income plus interest on equity capital. Net income was estimated to be $10,000 per dairy farm in the watershed in 1973.

Table 5.8. Estimated cost of reducing phosphorus loss from farming in Fall Creek watershed, Model II.

% Reduction in P	% Reduction in Corn	% Reduction in Cows	Cost to Watershed	Cost per Farm	% of Net Income
10	0	0	$ 0	$ 0	0
20	4	0	31,600	243	0
30	16	0	76,500	588	6
40	29	0	136,600	1,051	11
50	45	0	201,100	1,547	15
60	67	0	278,300	2,141	21
70	91	0	434,000	3,338	33
80	100	23	1,114,700	8,575	86
81	100	21	1,256,000	9,662	97

opportunity to leave land idle allowed a greater reduction of phosphorus loss (81%) to be achieved, because idle land had lower losses than land in hay. However, the reduction in phosphorus loss beyond 60% was very costly because feed was purchased to compensate for idle land.

Model III

In the third model, conservation practices were introduced. On appropriate soil types and slopes, contour strip-cropping was used

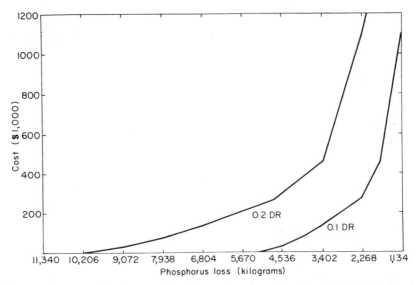

Figure 5.3. Cost of reducing phosphorus loss from farming in the Fall Creek watershed—Model II.

as a conservation practice to reduce soil and nutrient losses while maintaining crop production. Therefore the need for feed purchased outside the watershed should be reduced.

Figure 5.4. Strip-cropping (not contour) in the Fall Creek watershed. Strips of corn are alternated with strips of alfalfa-grass hay. (Photo by G. Casler)

Introduction of contour strip-cropping allowed each level of phosphorus restriction to be achieved at lower cost to the farmers in the watershed (Figure 5.5) than in Model II. No costs were attached to installation of contour strip-cropping. Assuming that there is at least some cost for installing and maintaining contour strip-cropping, the costs in Table 5.9 and Figure 5.5 are understated. However, the difference between the costs in Figure 5.3 and 5.5 or Tables 5.8 and 5.9 could be considered the maximum annual amount that farmers could afford to pay for strip-cropping and yet be better off than achieving the same phosphorus restriction without this conservation practice. For example, the cost of achieving a 40% reduction in phosphorus is $96,400 less with than without conservation practices. This is $11.69 per ha for the 8236 ha of cropland in the watershed. For the 3707 ha actually put in strip-cropping, the difference in cost is $26 per ha. The $26 is the maximum annual cost per ha that farmers could afford to pay for contour strip-cropping to achieve a 40% reduction in phosphorus and still be better

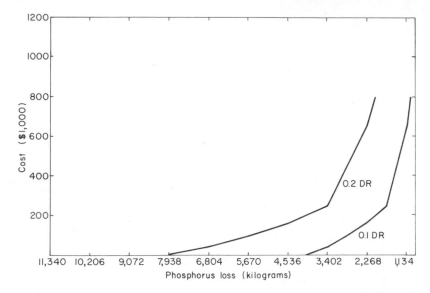

Figure 5.5. Cost of reducing phosphorus loss from farming in the Fall Creek watershed—Model III.

off than by achieving the same reduction entirely from changing corn to hay as in Model II. Corn production is reduced by only 3% compared to the 29% required to achieve the 40% reduction in phosphorus loss in Model II.

Although there is currently some strip-cropping practiced in the watershed, most of it is not actual contour strip-cropping and may not be very effective in reducing erosion and nutrient losses. To the extent that it is effective, the impact of the contour strip-crop-

Table 5.9. Estimated cost of reducing phosphorus loss from farming in Fall Creek watershed, Model III.

% Reduction in P	% Reduction in Corn	Cost to Farming	Cost per Farm	% of Net Income
10	0	$ 0	$ 0	0
20	0	0	0	0
30	0	300	2	—
40	3	40,200	309	3
50	14	94,700	728	7
60	27	160,500	1,235	12
70	51	245,000	1,885	19
80	67	656,500	5,050	51
82	69	795,500	6,119	62

ping considered in the model is overestimated because the initial solution assumed no contouring.

Model IV

In Models I, II and III, the reduction in phosphorus loss from agriculture is achieved by reducing the production of feed and replacements in the watershed and purchasing these production inputs from outside the watershed. To the extent that production of feed and replacements causes nutrient losses in other watersheds, the reduction of losses in Fall Creek results in a transfer of losses to other watersheds. To reduce such a transfer, Model IV allowed no increase in purchase of feed and replacements from outside the watershed above the levels in the initial solution. In that solution, no replacements and no hay but 7956 mt of grain (4960 mt of 16% dairy ration and 2996 mt of 40% protein supplement) were purchased from outside the watershed which approximates the situation existing in the watershed in 1973. In Model IV, 7956 mt of grain, any combination of 16% and 40% protein, was allowed to be purchased.

The restriction on purchase of inputs from outside the watershed raised the cost of achieving each level of reduction in phosphorus loss from agriculture. (Compare Figures 5.5 and 5.6, Tables 5.9 and 5.10.) Except for the last 2% reduction, the cost increases were relatively small, suggesting that increased purchases of feed and replacements from outside the watershed as a method of reducing phosphorus losses in the watershed was only minimally profitable at the cost-price relationship used in the model.

Table 5.10. Estimated cost of reducing phosphorus loss from farming in Fall Creek watershed, Model IV.

% Reduction in P	% Reduction in Corn	% Reduction in Cows	Cost to Watershed	Cost per Farm	% of Net Income
10	0	0	$ 0	$ 0	0
20	0	0	0	0	0
30	1	0	0	0	0
40	3	3	47,600	366	4
50	13	8	130,400	1,003	10
60	23	15	247,600	1,905	19
70	43	25	397,100	3,055	31
80	67	42	700,500	5,388	54
82	69	53	1,030,600	7,927	79

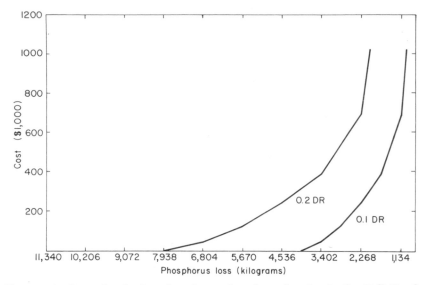

Figure 5.6. Cost of reducing phosphorus loss from farming in the Fall Creek
watershed—Model IV.

The restrictions on purchased inputs led to a substantial reduction in milk production in the watershed. For example, at the 60%
reduction in phosphorus loss, cow numbers were maintained at the
original level of 6800 in Models I, II and III but declined to 5781 in
Model IV. This implies a reduction of nearly 5 million kg of milk
sales (15%) in the watershed. If this milk production is needed by
consumers and is produced in other watersheds, then nutrient
losses from agriculture in those watersheds may be increased.

Biologically Available Phosphorus

The costs of reducing phosphorus loss from agriculture in the
watershed must be considered in relation to benefits achieved. The
benefits of reduced levels of phosphorus will come largely from
reduced algal production in Cayuga Lake. Because algal growth is
related to the fraction of phosphorus that is biologically available,
costs should be considered in relation to the reduction in BAP
leaving the watershed.

The results of the models suggest that small reductions in phosphorus losses from farming in the watershed could be achieved at
relatively low cost. Larger reductions, particularly in relation to
the decrease in BAP, become increasingly expensive.

Assuming that a 0.1 delivery ratio and a 0.2 BAP/P ratio are reasonable approximations of the Fall Creek situation, each 10% reduction in phosphorus loss implies a reduction of 114 kg of BAP. In Model II the average cost per kg reduction in BAP ranged from $139 for a 20% reduction to $1367 for an 81% reduction. The marginal cost of the last 1% reduction in phosphorus loss (80% to 81%) is $12,460 per kg of BAP. If the BAP/phosphorus ratio was 0.1 rather than 0.2, a 10% reduction in BAP would be 57 kg and each of the preceding cost per kg estimates would be doubled.

Limitations

In all four models the cost of achieving a given level of decrease in phosphorus loss associated with soil loss via land runoff is highly dependent on the physical and cost coefficients used in the model. Three of the most important of these are the crop yield levels, the cost of purchased feed, and the loss coefficients for biologically available phosphorus. The impact of each of these coefficients on cost of reducing phosphorus losses is discussed below.

The restrictions on phosphorus losses cause hay acreage, which presumably has lower phosphorus losses, to be substituted for corn acreage. While there is general agreement that corn silage will produce more digestible energy per acre than will hay crops in areas such as Fall Creek, there is less agreement on the actual difference in energy. In the Fall Creek model, yield levels were based largely on data from the Tompkins County Soil Survey. For example, a soil which could produce a 31.4 mt (9.4 mt dry matter) corn silage yield was assumed to produce a 7.8 mt (6.8 mt dry matter) legume hay yield under similar management levels (see Tables 5.2 and 5.3(a)). This yield level was used as the average of the first three years of a hay stand. The fourth and later years were credited with a yield level of 4.5 mt (4.0 mt dry matter) of grass hay. All rotations in the model were seven years. When hay was substituted for corn, in effect an acre of 31.4 mt corn silage was replaced by an acre of 4.5 mt grass hay. While the crop yields used in the model were believed to closely resemble those found in the watershed, it may be that if farmers were required to shift from corn to hay, that hay yields would be higher relative to corn than currently appears to be the case in Fall Creek. The costs presented in Tables 5.7–5.10 and Figures 5.2, 5.3, 5.5 and 5.6 for achieving the phosphorus restrictions in the model assume that corn would be replaced by higher yielding predominately legume (e.g., 7.8 mt) rather than by predominantly grass hay. Model results were adjusted to conform with this assumption. This implies improved management of legume stands or shorter rotations. Unless management of hay crops relative to corn crops was improved substantially, the estimated costs of meeting the restrictions probably would have been minimum costs. It is doubtful that these cost levels would be achieved. One concern is that as hay acreage is increased, the difficulty of harvesting at the period of high quality is increased. Thus average quality may decline.

In all four models, 16% dairy feed was priced at $132 per mt while 40% dairy supplement was priced at $165 per mt. These were the prevailing prices in the early summer of 1974. Fall 1974 prices were closer to $165 and $193 per mt. If these prices had been used in the model, the cost to farmers of reducing phosphorus losses in Model I, II, and III would have been higher. For example, in Model I and a 40% phosphorus reduction the cost would have increased from $148,000 to approximately $242,000, an increase of 64%.

Another limitation of the cost estimates for achieving reductions in phosphorus losses related to soil erosion is the uncertainty relative to predicted losses of BAP from various cropping alternatives. Although there is research-based evidence that soil and particulate phosphorus losses are greater from corn than from hay crops grown on the same soil type, there is little research-based evidence that BAP in surface runoff, per unit of land area, is different for corn than for hay crops grown on the same soil type. Our models assume that BAP losses are greater from corn. Until research data is available to support this assumption, the costs estimated by our model must be considered to be tentative. The question that needs to be answered is whether the universal soil loss equation and phosphorus enrichment ratios are the appropriate method to estimate BAP or TSP in surface runoff to streams from farm land.

LAND RUNOFF AS RELATED TO MANURE APPLICATION

In addition to the concern over runoff losses of soil and nutrients from cropped fields, livestock manure has received attention as a potential source of nutrients in runoff. The emphasis has been on the loss of nutrients from manure spread on frozen ground during the winter months.

This section provides information on the economic, physical and environmental trade-offs between alternative dairy manure han-

Figure 5.7. The picture on the bottom shows the pattern of manure spreading during the winter on a field in the Fall Creek watershed. The picture on the top is a closeup of manure spreading on a snow-covered field. (Photos by R. Oglesby and C. Ostrander, respectively)

dling systems. The economic evaluation considers only the manure handling aspects of the dairy operation. For the environmental evaluation, each alternative manure handling system was ranked for each of seven different environmental characteristics. This ranking was based on a questionnaire sent to 16 individuals familiar with alternative dairy farm manure handling systems. This method of evaluating the environmental implications was used because of the subjective nature of some of the environmental characteristics, such as odor and appearance. Furthermore, there is very little data available regarding the environmental effects of alternative manure handling systems.

Description of Manure Handling Systems

Manure handling methods examined consisted of stanchion housing with conventional or liquid systems and free-stall housing with conventional, liquid, or lagoon systems (Table 5.11). Each system considered was divided into three components: (1) collection, (2) storage, and (3) disposal.

A brief description of the storage facilities is needed to clarify the type of structure being considered. Stacking refers to a roofless manure stacking operation where the base and walls are made of concrete. With this type of storage, manure can be handled with conventional equipment. Two types of liquid storage structures are considered. The first is a tank built of concrete below ground level with a concrete top. The second is an earthen lagoon which may be lined, depending on the type of soil, and requires an underground pump to transfer the manure to storage.

Investment and Annual Costs of Disposal Systems

Investment and annual costs are largely a function of the type of housing, manure handling equipment, type of storage facility, length of storage period, and herd size. Investments required for three herd sizes, two types of housing and the alternative manure systems are given in Table 5.12. In all cases, six months' storage is provided.

In calculating the storage costs, the following assumptions with respect to storage space and construction costs were made:

1. Manure stacking: 0.071 m^3/cow/day @ \$10.60/m^3
2. Liquid storage (below grade): 0.074 m^3/cow/day @ \$30/m^3

Figure 5.8. A manure stacking system. The manure is loaded into the storage area with a mechanical conveyor and removed with a loader mounted on a tractor. The walls around the storage are constructed to control runoff from the storage. (Photo by R. Guest)

For each system, the investment includes the storage and manure handling equipment required by that system. Annual costs per cow for equipment, storage and electricity and labor and tractor hours required by each type of manure handling system, herd size, and length of storage are given in Table 5.12. Annual costs of each manure handling system include depreciation, interest, real estate taxes, and insurance as well as variable costs such as repairs and fuel which are a function of the amount of use made of the equipment. Annual costs also include the labor and tractor hours spent annually for cleaning stalls, scraping, loading, agitating, pumping and hauling manure under alternative systems (Jacobs and Casler, 1974).

Changing from daily spreading to a storage system will reduce the labor and tractor hours required. This decreased time is due to elimination of such daily tasks as hitching and unhitching the spreader plus the additional spreading time for partial loads. Also, larger spreaders are normally used with storage systems. A large decrease in labor and tractor hours also occurs in free-stall systems with the use of a mechanical rather than a tractor scraper. Labor

Table 5.11. Summary by components of alternative dairy manure handling systems.

Collection	Storage	Disposal
Stanchion Housing:		
1. Gutter cleaner	—	Spreader (daily)
2. Gutter cleaner-stacker	Stacking	Spreader (spring and fall)
3. Gutter flush	Liquid	Liquid spreader (spring and fall)
4. Gutter flush	Liquid	Liquid spreader and soil injector
Free-Stall Housing:		
1. Tractor scraper-manure lip	—	Spreader (daily)
2. Mechanical scraper	—	Spreader (daily)
3. Tractor scraper-stacker	Stacking	Spreader (spring and fall)
4. Mechanical scraper-stacker	Stacking	Spreader (spring and fall)
5. Mechanical scraper	Liquid	Liquid spreader (spring and fall)
6. Mechanical scraper	Liquid	Liquid spreader with soil injector (spring and fall)
7. Mechanical scraper	Aerated lagoon	Irrigation system (spring, summer, and fall)
8. Tractor scraper	Liquid	Liquid spreader (spring and fall)
9. Tractor scraper	Liquid	Liquid spreader with soil injector (spring and fall)
10. Tractor scraper	Aerated lagoon	Irrigation system (spring, summer, and fall)

and tractor hours per cow decrease as herd size increases. Even though the labor requirement is reduced by changing from daily spreading to storage, the labor requirement of storage systems is concentrated in the spring where it conflicts with labor needed for cropping. Thus the reduction in labor costs due to storage probably overstates the actual reduction in cost to the farm business.

A summary of total investment and cost/cow/year by herd size and type of manure handling system is presented in Table 5.12. While the annual manure handling cost per cow for any system tends to decline with increased herd size, the largest difference in manure handling costs are associated with changes from daily

Figure 5.9. A liquid manure pump transferring manure from an underground storage tank to a liquid manure spreader. (Photo by C. Ostrander)

spreading to storage. For example, with 100-cow herds, either stanchion or free-stall, annual cost per cow for daily spreading systems is about $40 per cow; stacking increases the cost to about $55–$60 per cow and liquid storage systems have costs in the $75 to $85 range. The increased costs largely result from the inability of labor and tractor savings to offset the increase in investment required by the storage systems. Research to find lower cost forms of acceptable storage might have a high payoff, both to farmers and society, if farmers are encouraged or forced to store manure for six months or more.

Minimization of the cost of manure handling is not the only consideration in planning such a system. Other reasons why a dairyman might want to handle the dairy manure on other than a daily basis are: (1) elimination of a disagreeable everyday job, (2) increased manure value, and (3) environmental concerns or pressures.

While storage would eliminate the disagreeable everyday job of spreading manure, it substitutes a possibly even more disagreeable job, due to odor problems, that must be concentrated at a period of time in the spring where it conflicts with land preparation and corn planting on most dairy farms.

There may be some increase in the value of manure from a stor-

Table 5.12. Summary of investment and annual costs per cow by herd size and manure handling systems.[1]

	50 Cows		100 Cows		200 Cows	
	Total Investment	Annual Cost per Cow	Total Investment	Annual Cost per Cow	Total Investment	Annual Cost per Cow
			Dollars			
Stanchion Housing:						
1. Gutter cleaner-daily	3,200	44	5,200	42	3,600	34
2. Cutter cleaner-stacking	13,000	69	21,800	59	7,100	38
3. Gutter flush-liquid	25,700	81	46,600	72	34,300	51
4. Gutter flush-liquid-soil injection	26,400	84	47,300	74	38,400	48
Free-Stall Housing:						
1. Tractor scraper-daily	3,300	45	3,300	38		
2. Mechanical scraper-daily	3,500	45	4,600	41		
3. Tractor scraping-stacking	12,600	68	19,300	56		
4. Mechanical scraper-stacking	13,400	62	21,200	52		
5. Mechanical scraper-liquid	25,800	83	46,800	75	91,300	72
6. Mechanical scraper-liquid-soil injection	26,500	87	47,500	78	92,700	75
7. Mechanical scraper-lagoon-irrigation	12,000	72	18,600	67	30,100	62
8. Tractor scraper-liquid	25,000	91	44,900	81	87,300	77
9. Tractor scraper-liquid-soil injection	25,700	95	45,600	83	88,700	79
10. Tractor scraper-lagoon-irrigation	11,200	78	16,700	72	26,000	66

[1] These costs were computed in 1973. Increased costs for storages, equipment, fuel, and labor since 1973 would increase the investment and annual costs for all systems. Systems with the larger investment would have the largest increases in investment and costs.

age system compared to manure from a daily spread system because of decreased losses of nitrogen, phosphorus, and potassium. However, the magnitude of the decreased loss through storage depends so heavily on the type of management that any estimate of the added manure value would be only one of many possible estimates that could be made. Furthermore, research not reported here indicates that the added value of stored manure would not offset the additional cost of storage except under unusual combinations of conditions not likely to occur.

Manure Storage in Fall Creek Watershed

The 1973 survey of farmers in the Fall Creek watershed indicated that little or no manure from milking herds is stored in the winter for spring application. The cost to farmers in the watershed of a regulation prohibiting winter spreading can be estimated for either stacking or underground liquid storage from the cost data reported above. The average milking herd size in the watershed is 52 cows. Therefore the cost estimates for 50-cow herds were used as the basis for cost estimates for the watershed. If all farmers installed stacking systems, the additional investment required would be approximately $10,000 per farm or a total of $1.3 million for the watershed. The added annual cost, at $25 per cow, would be $170,000 per year or $1310 per farm. For liquid systems with below ground storage, the added investment is $22,000 per farm or $2.9 million for the watershed. The added annual cost, at $45 per cow, would be $306,000 for the watershed or $2350 per farm. These annual costs would be reduced by any additional value of manure due to storage.

The cost estimates presented above are for manure from the milking herd only. Manure storage for replacement animals would add about 30% to these costs. The 30% increase to provide storage for manure from replacements results in the following estimates for the watershed. For stacking systems, the investment would be $13,000 per farm, or $1.7 million for the watershed. The annual cost would be $221,000 for the watershed or $1700 per farm. For liquid systems, the investment would be $28,600 per farm or $3.7 million for the watershed. The annual cost would be $397,800 for the watershed or $3060 per farm.

Environmental Considerations

Having estimated the added cost of manure storage, the question arises: "What are the benefits derived from this expenditure?"

There is no widely accepted definition of the term "environmental quality," or of what constitutes a "benefit." Increasing awareness and concern about the quality of the environment has brought all sources of pollutants under suspicion, including livestock manure. Therefore, an analysis was made of the potential environmental impact of alternative dairy manure handling systems.

In the analysis, the environmental characteristics considered were noise, appearance, flies, odor, loss of nutrients, and risk to water quality. While some of these characteristics, such as nutrient loss, are measurable, there is little research data currently available. Furthermore, there is at present no objective way to measure characteristics like odor and appearance.

Therefore, to obtain a measure of the environmental impact of alternative dairy manure handling systems, an opinion survey was used to develop environmental impact scores (Jacobs and Casler, 1972). A summary of the weights placed on characteristics, the weighted scores of characteristics and the environmental impact scores for alternative manure handling systems are presented in Table 5.13. The higher the environmental impact score, the greater the environmental degradation.

Several observations can be made from comparing impact scores of individual environmental characteristics by systems and impact scores for the various manure handling systems. One of the most interesting comparisons is between the characteristics "odor" and "risk to water quality." Table 5.13 points out that the characteristic "risk to water quality" received the highest score for daily spread systems and the characteristic "odor" received the highest score for systems with manure storage. The important point is to realize the trade-offs being made among environmental characteristics common to alternative systems. For example, if one was concerned only with water quality, it might be concluded that manure storage would be the answer. However, if one considers the impact of manure storage on other environmental characteristics, the above conclusion may no longer be so obvious.

Effect of Manure Storage on Nutrient Runoff

Much of the environmental concern about manure handling is related to nutrient runoff from winter spreading, particularly on frozen ground. Experimental results have shown that large nutrient runoff can occur when winter spreading of manure was followed by a rather large rainfall or snowmelt event (Klausner and Zwerman,

Table 5.13. Summary of environmental impacts for selected stanchion and free-stall manure handling systems.

	Odor	Flies	Appearance	Noise	Loss of Nutrients Storage	Loss of Nutrients Land Disposal	Risk to Water Quality	Estimated Environmental Impact Score[1]
Weight of characteristics	8.5	6.6	4.8	5.3	4.1	4.9	8.1	
				Weighted Value of Characteristics				
Stanchion Housing:								
1. Gutter cleaner-daily	36	27	19	15	4[2]	29	49	179
2. Gutter cleaner-stacking	47	41	24	16	13	20	32	193
3. Gutter flush-liquid	65	28	19	13	11	20	31	187
4. Gutter flush-liquid-soil injection	48	24	16	13	10	14	20	145
Free-Stall Housing:								
1. Tractor scraper-daily	37	32	22	15	4[2]	28	46	184
2. Mechanical scraper-daily	32	29	17	15	4[2]	27	45	169
3. Tractor scraper-stacking	49	44	30	16	14	18	27	198
4. Mechanical scraper-stacking	44	40	30	16	16	22	26	198
5. Mechanical scraper-liquid	65	29	18	15	10	20	28	185
6. Mechanical scraper-liquid-soil injection	48	24	14	15	9	18	23	151
7. Mechanical scraper-lagoon-irrigation	34	26	22	14	17	19	27	159
8. Tractor scraper-liquid	64	28	18	15	8	20	29	182
9. Tractor scraper-liquid-soil injection	51	25	13	15	10	17	23	154
10. Tractor scraper-lagoon-irrigation	35	28	20	13	18	18	28	160

[1] The environmental impact score is the sum of the weighted values of the characteristics.
[2] Since daily spreading involves no storage, this represents the minimum value for that characteristic.

1974; Minshall, *et al.*, 1970). However, the experimental results cited were collected from experiments where the entire plot was covered with manure in one day. These experiments do not simulate daily spreading as typically practiced by farmers. Therefore, any estimates of the effect of manure handling practices such as storage and time of application on nutrient runoff losses must be considered tentative and subject to modification as additional data become available.

Data on the runoff losses as related to dairy manure handling practices in northern climates have recently been collected in Wisconsin and New York. Some results from each experiment and from our manure handling cost study were combined in an effort to compute costs of reducing phosphorus losses to Fall Creek.

In the three-year Wisconsin study (Minshall, *et al.*, 1970) conducted at Lancaster, manure was spread on plots with 10 to 12 percent slope. Runoff and nutrient losses were measured from plots with no manure, fresh manure applied in the winter and fermented and liquid manure applied in the spring. Corn was grown on all plots each year. Annual nutrient losses from each treatment are shown in Table 5.14. In the first year, nitrogen and phosphorus runoff losses were five to six times higher on the winter spread plots. In the second year, nitrogen and phosphorus losses were about equal for all three treatments. In the third year, winter spreading

Table 5.14. Annual nutrient losses from amanured and nonmanured corn plots, 1967–69, Lancaster, Wisconsin.

		Manure treatment		
Nutrient	None	Fresh Winter	Fermented Spring	Liquid Spring
		kg per hectare		
1966–67				
Total N	5.54	26.90	5.32	5.08
Total P	1.22	5.77	0.96	1.18
1967–68				
Total N	3.61	3.05	3.35	2.88
Total P	1.49	1.03	0.75	0.95
1968–69				
Total N	3.95	8.04	3.38	2.81
Total P	1.22	2.01	0.68	0.76
Average				
Total N	4.36	12.67	4.02	3.59
Total P	1.31	2.94	0.81	0.96

Source: Minshall, Witzel, and Nichols, 1970.

resulted in nitrogen and phosphorus losses two to three times that of spring spreading. Average losses for the three years were considerably higher on the winter spread plots. While the average nutrient losses of these three years may not represent average conditions, some calculations follow based on the assumption that they do.

A comparison was made between losses from winter and spring applications of manure. Losses from the two spring treatments were similar. Therefore, these two treatments were averaged and designated "spring." Runoff losses from winter spreading were 8.87 and 2.06 kg per ha greater for nitrogen and phosphorus, respectively, than from spring spreading.

A study of nutrient runoff was conducted at the Cornell Agronomy Research farm near Aurora, N.Y. (Klausner and Zwerman, 1974). Nutrient losses were measured from plots on which manure was applied in the winter, spring, and summer. Annual losses of inorganic nitrogen and total soluble phosphorus averaged over the calendar years 1972, 1973, and 1974 and good and poor soil management for the 35 mt/ha (15 t/A) rate for winter and spring applications are shown in Table 5.15. Average inorganic nitrogen losses were 40% higher and soluble phosphorus losses 255% higher from winter than from spring spreading. Losses of phosphorus were 0.28 kg/ha higher with winter spreading.

Because nitrogen levels in Fall Creek are far below the U.S. Public Health Service standard at all times, nitrogen losses were judged

Table 5.15. Annual runoff losses of inorganic nitrogen and soluble phosphorus from manure applied at 35 mt/ha in winter and spring, Aurora, N.Y., 1972, 1973, and 1974, average of good and poor soil management.

		Time of Application	
		Winter	Spring
Year	Nutrient		
		kg/ha	
1972	N	5.33	3.84
	P	0.66	0.22
1973	N	1.10	0.62
	P	0.41	0.05
1974	N	1.22	1.00
	P	0.11	.07
Average 1972–74	N	2.55	1.82
	P	0.39	0.11

Source: 1972 and 1973 data from Klausner and Zwerman (1974); 1974 data from personal communication from S. D. Klausner.

not to be an environmental problem. Therefore, the cost of reducing nutrient losses by storage to avoid winter spreading of manure can logically be charged to the reduction of phosphorus losses.

The next step was to calculate the cost per kg of reduced phosphorus loss if manure is stored and spring spread rather than spread in the winter. As stated earlier, the annual cost of manure storage in the Fall Creek watershed would be approximately $398,000 for liquid systems and $221,000 for stacking systems, less any increase in manure value due to avoidance of winter spreading. Most of the manure spread in winter is applied to the land to be used for the 2760 ha of corn. Much of this land has a lower potential for runoff than the land used in the Wisconsin and New York studies, but some has a higher runoff potential. Considering this situation, calculations were made based on the assumption that runoff losses of phosphorus on the 2760 ha of corn land could be reduced by the differences between winter and spring spreading found in either the Wisconsin or New York studies if winter storage of manure was practiced in the watershed. A delivery ratio of 0.5 was assumed, which is greater than the soil delivery ratio but used because the delivery ratio could be much higher for phosphorus from manure sometimes applied on frozen ground and subject to winter and early spring runoff. It is recognized that all these assumptions are very tenuous.

Based on the Wisconsin study and a delivery ratio of 0.5, the cost per kg reduction in total phosphorus loss is $140 for liquid storage and $78 for stacking. Not all this phosphorus is soluble, therefore the cost per kg of soluble phosphorus would be greater.

In the New York study, with losses of soluble phosphorus 0.28 kg per ha higher from winter spreading and a delivery ratio of 0.5, the reduction in soluble phosphorus from the watershed would be 385 kg per year. The cost per kg reduction in soluble phosphorus is $1034 for liquid storage and $574 for stacking. The costs computed from the New York and Wisconsin data are not comparable because one is based on total and the second on soluble phosphorus.

In either case, variations in the delivery ratio could substantially alter the cost per kg reduction in phosphorus loss to Cayuga Lake.

Summary of Manure Handling Systems

Results of the above analysis indicate that manure handling systems which include storage would tend to decrease nutrient losses to water. These same systems would also increase the cost of handling dairy manure. Furthermore, the environmental analysis section points out that storage systems, without soil-injection or aeration, not only have a higher cost but higher environmental impact scores. The reason is that while storage systems may reduce the adverse environmental effects related to nutrient runoff from winter spreading, they tend to increase other environmental effects such as odor and flies. A summary of the costs and environmental impact scores for alternative dairy manure handling systems is presented in Table 5.16.

The above analysis is an illustration of the possible trade-offs among environmental characteristics. Because of these trade-offs, the relevant environmental characteristics must be included to obtain the overall environmental impact of the alternatives being considered. This suggests that recommendations for winter storage of manure as a means to control runoff of nutrients could create other environmental impacts as bad or even worse than the runoff problem. It also indicates the need to consider the major environmental characteristics of alternative waste handling systems before making recommendations.

BARNYARD RUNOFF

On September 7, 1973, notice was published in the Federal Register that the Environmental Protection Agency (EPA) was proposing effluent limitation guidelines for the feedlots category of point sources. At present, these regulations are focused on large feedlots and for dairy operations this is defined as 700 or more dairy cattle. For smaller operations, the EPA is reviewing information submitted during the public comment period to evaluate the economic impact of regulations for smaller herds.

The purpose of this section is to estimate the economic impact and reduction in runoff losses of phosphorus from barnyards if the effluent guidelines are imposed on New York dairy farmers. The extent of these impacts depends on a number of factors including: (1) size of the production unit; (2) site, soils and topography of barnyard area; (3) size of storm for which runoff is to be controlled; (4) length of storage period; and (5) number of production units affected by the regulation. To obtain information on these factors, a one-page questionnaire was attached to the report sheets filled out each year by Dairy Farm Business Management Cooperators. In 1973, 609 farmers submitted records, and 381 returned the barnyard runoff survey, 358 of which were complete enough to be used in the analysis.

By combining the farm business records and the runoff survey,

Table 5.16. Comparison of costs and environmental impact scores for selected dairy waste handling systems.

	Total Investment[1]	Annual Cost per Cow[1]	Environmental Impact Score[2]
	$	$	
Stanchion Housing:			
1. Gutter cleaner-daily	5,200	42	179
2. Gutter cleaner-stacking	21,800	59	193
3. Gutter flush-liquid	46,600	72	187
4. Gutter flush-liquid-soil injection	47,300	74	145
Free-Stall Housing:			
1. Tractor scraper-daily	3,300	38	184
2. Mechanical scraper-daily	4,600	41	169
3. Tractor scraper-stacking	19,300	56	198
4. Mechanical scraper-stacking	21,200	52	198
5. Mechanical scraper-liquid	46,800	75	185
6. Mechanical scraper-liquid-soil injection	47,500	78	151
7. Mechanical scraper-lagoon-irrigation	18,600	67	159
8. Tractor scraper-liquid	44,900	81	182
9. Tractor scraper-liquid-soil injection	45,600	83	154
10. Tractor scraper-lagoon-irrigation	16,700	72	160

[1] These cost estimates are based on a 100-cow herd and come from Table 5.12.
[2] The environmental impact scores are taken from Table 5.13.

data was obtained on: (1) distribution of farms for the specified size groups; (2) barnyard area per cow for the specified size groups; (3) location of the barnyard relative to a stream or road ditch; and (4) number of farms that have a barnyard. Using this information plus rainfall and cost data, the costs of constructing runoff control facilities for three herd sizes were computed. The remainder of this section presents the specific assumptions and data underlying the estimated cost of runoff control facilities and the associated reduction in phosphorus losses.

Effluent Guidelines for Feedlots

An evaluation of the effluent limitation guidelines for feedlots requires a definition of feedlots. The EPA has defined a feedlot as "a concentrated, confined animal or poultry growing operation for meat, milk or egg production, or stabling, in pens or houses wherein the animals or poultry are fed at the place of confinement and crop or forage growth or production is not sustained in the area of confinement" (Federal Register, 1973). Effectively, a feedlot is any confined area where animal density precludes vegetative cover.

In addition to the feedlot definition, the degree of effluent reduction to be obtained must be specified. The effluent limitations guidelines specified by EPA are comprised of two different levels of regulation and effective dates. One, to take effect July, 1977 specifies "no discharge of process wastewater pollutants[1] to navigable waters except for runoff which is not contained by facilities, designed, constructed and operated to contain all process wastewater in addition to the runoff from the 10-year, 24-hour rainfall event as established by the U.S. Weather Bureau for the region in which the point source discharger is located," (Federal Register, 1973). The second, to take effect July, 1983 is identical except that the rainfall event is changed to a 25-year, 24-hour event.

In addition to these guidelines, the Department of Environmental Conservation in New York State and the EPA have taken the position that spray irrigation of waste water requires a permit. The permit is an attempt to assure that the discharge of wastewater on a spray irrigation field meets the following conditions:

1. Spray irrigation shall be practiced only during the period of May 1 to November 1,

---

[1] The definition of "process wastewater" includes any precipitation which comes into contact with any manure.

2. Spray irrigation shall be practiced during daylight hours with no spraying during periods of precipitation,
3. No area of spray field shall be irrigated on two consecutive days, and
4. Surface runoff of irrigated waste water from the spray field shall not be permitted.

Whether or not dairy farms would have to obtain such a permit is uncertain at present. However, if the area and characteristics of a feedlot and the appropriate regulations are known, a system to accommodate the regulations and the costs associated with it can be estimated.

Runoff Control Facilities

After reviewing runoff and spray irrigation regulations, it was decided that the runoff control facility should be constructed to hold the runoff for the 6-month period from November 1 to May 1 plus the 25-year, 24-hour event. It was estimated that runoff was approximately 50% of the precipitation during this 6-month period. For the 25-year, 24-hour event, runoff was estimated to be equal to precipitation. The estimated runoff, including the 25-year, 24-hour event, is 31.5 cm. The computed runoff times the area of the barnyard yields an estimate of the volume of storage required for lot runoff.

The investment and cost of constructing and operating a facility for controlling this volume of runoff can now be computed. The type of facility under consideration is shown in Figure 5.10. It consists of: (1) a diversion terrace on three sides of the barnyard; (2) a spillway; (3) a holding pond with a fence; and (4) an irrigation system.

The investment and annual costs of constructing and operating such a system for 40-, 80-, and 120-cow dairies are presented in Table 5.17. Investment included pond excavation at $0.92 per cubic meter of storage volume, diversion terrace at $1.60 per meter and fencing at $1.51 per meter.* Pond excavation investment includes provision of additional volume of sedimentation of solids. Sedimentation was estimated to be approximately 227 m^3 per year for a 1-ha watershed.** Investment for the irrigation equipment consists of a

---

* The fencing cost includes $.39 per m for wire, $.46 per m for posts and $.66 per m for labor for a 4-strand barbed wire fence.
** This estimate was obtained from a personal communication with Richard Crowe (1974), who is an area engineer with SCS. He indicated the estimate was obtained using the soil loss equation for a slope with a gradient of 3 percent and a length of 61 m under fallow conditions.

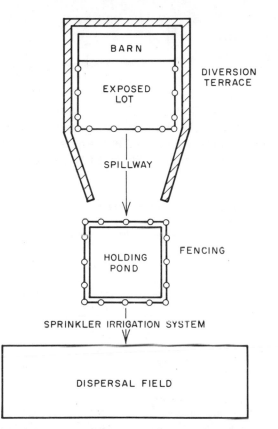

Figure 5.10. Facility to control barnyard runoff.

5 H.P. electric motor at $800, a manure gun at $275 and 7.62 cm aluminum pipe at $3.28 per meter.

To illustrate how the costs in Table 5.17 are calculated, the 80-cow herd is used as an example:

The feedlot area is 42.7 m²/cow × 80 cows = 3416 m² = 0.34 ha
Runoff   = 0.31 m × 0.34 ha = 0.1054 ha m = 1067 m³
Sedimentation = 76 m³/yr × 15 yr   = 1140 m³
     Total Storage Volume   2207 m³

Using a pond depth of 2.44 m with a side slope of 3:1 and 0.61 m of freeboard, the cubic meters needed and the dimensions can be calculated with the following formula for volume:

$$\text{Vol} = 1/3\text{h}[\text{A} + \text{a} + \sqrt{\text{Aa}}]$$

where,
  A = surface area of the pond
  a = area of the base
  h = depth

For the 2.44 m depth, $A = 1,546$ m^2, and $a = 610$ m^2 will provide 2,544 m^3 of storage. The surface is 39.3 $\times$ 39.3 m and the base is 24.7 $\times$ 24.7 m. To obtain the dimensions and storage for the additional 0.61 m of freeboard add 3.7 m to the surface dimensions and solve. Thus the final dimension is 43 $\times$ 43 m and a volume of 3905 m^3.

Table 5.17. Estimated cost of controlling barnyard runoff for 40-, 80-, and 120-cow dairies.[1]

	Dollars		
	40 Cows	80 Cows	120 Cows
Investment			
Pond[2]	1780	3270	5010
Diversion	170	210	250
Fence	204	260	310
Irrigation	2075	2575	2575
Total Investment	4229	6315	8145
Investment/cow	106	79	68
Annual Costs			
Ownership & Maintenance:			
Pond Diversion Fence[3]	240	410	615
Irrigation Equip.[4]	490	605	605
Labor[5]	115	125	135
Total Annual Cost	845	1140	1355
Cost/cow/yr	21	14	11

[1] A feedlot area of 42.7 meters per cow was used for all three herd sizes.
[2] The pond is 2.44 meters deep with 0.61 m of freeboard and the side slopes are 3:1.
[3] 15-year life is assumed with a 9% interest rate.
[4] 23.5% of the investment.
[5] 10 hours of labor is assumed for each set-up and tear-down plus 1 hour per 24 hours of operation.

Using this dimension and volume the investment amounts to $6,315 and the annual costs are $1,140. This translates to $2.87/mt of milk ($.13/cwt) for the 80-cow farm assuming milk production is 4990 kg (11,000 lb)/cow/yr.

The costs of controlling barnyard runoff ranged from $4.19/mt

($0.19/cwt) for a 40-cow dairy down to $2.20/mt ($.10/cwt) for a 120-cow dairy. Thus the runoff guidelines have a greater impact on the smaller dairy farms.

To expand these runoff control costs on individual farms to all dairy farms in New York State, information on the number of dairies affected and their distribution by herd size is needed. The barnyard survey previously mentioned indicated that 31% of the barnyards were located adjacent to a stream or road-ditch. Of this 31%, 12.5% were farms with less than 50 cows, 16.0% were farms with 50–99 cows and 2.4% were farms with 100 or more cows. Of the 900,000 dairy cows in New York, 380,000 are in herds of less than 50 cows, 350,000 in herds with 50–99 cows, and 170,000 in herds of 100 or more cows (Conneman, 1974). If only the 31% of farms adjacent to streams need runoff control facilities and the cost is spread over all milk produced in the state, the cost would be approximately $.88/mt ($0.04 per cwt) of milk sold.

An additional 45% of the farms surveyed had barnyards within 457 m of a stream. Some of these barnyards probably need runoff control facilities. To extend runoff control to one-third of this 45% would increase the cost to $1.32/mt ($0.06 per cwt); for two-thirds the cost would be $1.76/mt ($0.08 per cwt) and for all 45%, the cost would be $2.20/mt ($0.10 per cwt) of milk produced in the state.

Barnyard Runoff in Fall Creek Watershed

The data developed for dairy barnyard runoff control in New York is generally applicable to the Fall Creek watershed. Assuming that approximately the same percentage (31%) of the barnyards are adjacent to a stream or road ditch, and that the cost of runoff control is $0.88/mt ($0.04 per cwt) the annual cost per farm in Fall Creek would be approximately $230. The total cost for the watershed would be approximately $30,000 per year. On the farms needing runoff control the cost would be approximately $2.87/mt ($0.13 per cwt) or $750 per farm.

If barnyards other than those adjacent to streams or road ditches need runoff control, costs would be higher. For example, if 1/3 of those not adjacent to a stream but within 457 m need runoff control, the annual cost to farmers in the watershed would be approximately $345 per farm per year or a total of $45,000.

No generally applicable estimates of nutrient losses from barnyards in the watershed are available. Therefore it is difficult to make estimates of runoff control costs in relation to decreases in nutrient losses. Any estimates are tentative and subject to revision as additional data becomes available.

Data presented in Chapter 3 for one barnyard in the watershed indicates that 25 kg of BAP was picked up by the stream from the barnyard during a one-year period. If the annual cost of runoff control for this barnyard was equal to the average cost per barnyard in the watershed ($750) and phosphorus loss was reduced to zero, the cost per kg of phosphorus removal would be $30. This

Figure 5.11. A dairy barnyard in the Fall Creek watershed. (Photo by R. Oglesby)

barnyard is used by about 75 replacement heifers who obtain their water supply directly from the creek and have access to the barnyard at all times. The typical milking herd in the watershed would be housed most of the day and would not obtain water from the stream. Thus the 25 kg of phosphorus loss is probably high for the typical milking herd barnyard located adjacent to a stream. To the extent that runoff losses of phosphorus are lower, the cost per kg of phosphorus reduction will be higher. Data collected from a barnyard in Ohio in which 60 steers were confined indicated an annual loss of 5 kg of phosphorus (Edwards, *et al.*, 1971).

COST OF PHOSPHORUS REMOVAL FROM SEWAGE

Data presented in Chapter 3 and Table 5.6 indicate that approximately one-fourth of the total soluble phosphorus leaving the Fall Creek watershed came from the Dryden sewage treatment plant. Technology exists with which to remove phosphorus from sewage by tertiary treatment but it rarely has been used in plants as small as the one at Dryden which treats less than 0.2 mgd.

Most cost estimates for phosphorus removal apply to plants with capacities of 1 mgd or more and indicate considerable economies of size. In addition there is considerable variability in cost estimates and in some cases, not all costs are included. One estimate of the cost of phosphorus removal from domestic sewage by tertiary treatment was based on an article published by Kumar and Clesceri (1973) which reported costs approximately the same as costs presented in a report prepared by Environmental Quality Systems Inc. (1973). These estimates assume that the influent phosphorus content is 10 mg/l, that 10% is removed by secondary treatment and that an additional 70% of the original phosphorus content is removed by tertiary treatment for a total of 80% removal. Costs are based on a 1 million gallon per day (3785 m^3/day) plant. Each 1 mg/l of phosphorus removed is equal to 3.78 kg of phosphorus per million gallons of sewage. Removal of 7 mg/l is equivalent to removing 26.46 kg of phosphorus per million gallons of sewage. Total costs of phosphorus removal, including capital plus operating are $75.80 per million gallons, using either alum or ferric chloride. Therefore, the cost per kg of phosphorus removed is $2.86.

In areas such as New York where phosphate detergents have been banned, the phosphorus content of sewage is considerably reduced from preban levels. Thus tertiary treatment will remove smaller amounts of phosphorus. Assuming that capital and operating costs are the same, a 40 percent reduction in phosphorus content of influent, and that chemical and sludge removal costs are reduced proportionately to the decrease in phosphorus content, cost per kg phosphorus removed would be $3.90 for the original 10 mg/l sewage with a post-ban content of 6 mg/l.

The Dryden sewage treatment plant treats less than 0.2 mgd; therefore, the cost estimate was multiplied by 2.5. Construction, operating and chemical costs have increased since the estimates of Kumar and Clesceri were made. An adjustment of 25% was made to reflect these cost increases resulting in a cost of $12 per kg of phosphorus removed.

Monti and Silbermann (1974) have estimated that the cost of removing 95% of the phosphorus (adjusted to January 1975 cost levels) from sewage with 20 mg/l phosphorus content is approximately 10¢ per thousand gallons, using two-stage tertiary treatment in a million gallon/day plant. Based on these data, the cost per kg of phosphorus removed is approximately $1.40. Multiplying by 2.5 to adjust for higher costs in smaller plants results in a cost of $3.50 per kg. It is not clear whether the cost of sludge removal is included in Monti and Silbermann's estimates.

Considering both the foregoing estimates and the likelihood that Dryden's sewage had approximately 10 mg/l phosphorus prior to the detergent ban, the cost of removing phosphorus from Dryden's sewage was estimated to be in the range of $10 to $15 per kg.

From a broader perspective, the phosphorus input to Cayuga Lake could be substantially reduced by tertiary treatment of the sewage from larger population centers such as the City of Ithaca and the village of Cayuga Heights, the effluent from which directly enters the lake. The Cayuga Heights plant, which also handles sewage from some non-village residents, installed tertiary treatment for phosphorus removal during 1975. In such cases, not only does the phosphorus directly enter the lake rather than being deposited several miles upstream, but the cost of phosphorus removal would be lower because of treatment in larger plants.

COMPARISON OF COSTS OF PHOSPHORUS REMOVAL

As stated earlier, any estimates of costs of reducing phosphorus loss from farming in Fall Creek must be considered to be tentative, largely because the amount of phosphorus loss from farming activities is uncertain. Table 5.18 summarizes the estimates of the reduction in biologically available phosphorus input that might be expected to occur from each of the control methods discussed earlier, together with estimated cost of achieving such control. The research reported above indicates that the cost of reducing phosphorus loss from farming is significantly higher than the cost of removing phosphorus from domestic sewage. The cost per kg of reduction in biologically active phosphorus loss as related to soil erosion by changing from corn to hay crops appears to be on the order of hundreds of dollars, except possibly for reductions as small as 10 to 20%. The cost of reducing phosphorus loss by storage to avoid winter spreading of manure appears to be in the range of

Table 5.18. Summary of estimated costs of reducing phosphorus inputs from Fall Creek to Cayuga Lake.

Method of Phosphorus Reduction	Reduction in Biological Available Phosphorus	Annual Cost to Watershed	Cost per kg of Phosphorus Reduction
	kg	$	$
Reduction in corn acreage	680[1]	278,000	409
Avoidance of winter spreading of manure			
Liquid storage	385[2]	398,000	1,034
Stacking	385[2]	221,000	574
Control of barnyard runoff	1500[3]	45,000	30
	300[4]	45,000	150
Tertiary treatment of Dryden sewage	1200[5]	14,400	12

[1] Model II, 60% reduction in phosphorus, 0.1 delivery ratio and 0.2 BAP/P ratio.

[2] Difference of 0.28 kg of soluble phosphorus loss between winter and spring spreading, based on New York data (Klausner and Zwerman, 1974) and a 0.5 delivery ratio. Any increase in manure value due to storage would decrease the costs shown.

[3] 25 kg per barnyard from 46% of the barnyards.

[4] 5 kg per barnyard from 46% of the barnyards.

[5] May 1973–April 1974 estimate of input from Dryden sewage treatment plant, Table 3.16.

$575 to $1000 per kg of phosphorus. Any additional fertilizer value of manure due to storage would reduce these costs. The cost of reducing phosphorus loss by barnyard runoff control may be on the order of $30 to $150 per kg. Thus it appears that barnyard runoff control may be the lowest cost method, among the three studied, of reducing phosphorus loss from farming in the Fall Creek watershed. This conclusion is tentative and subject to modification with further research.

A comparison of the estimated cost of reducing phosphorus losses from farming with the cost of phosphorus removal from sewage indicates that the latter is much less costly. Of the three methods of reducing phosphorus losses from farming, barnyard runoff control is the only method that even approaches sewage treatment in terms of a low-cost method of phosphorus removal.

The data in Table 5.18 indicate that if all the control measures for reducing phosphorus losses from farming were undertaken, the reduction in biologically available phosphorus loss from farming in the Fall Creek watershed to Cayuga Lake would be 1365 to 2565 kg per year. This is 1.25 to 2.3 times the estimated average annual loading of total soluble phosphorus from all diffuse sources for two one-year periods indicating that some or all of the estimates of phosphorus loss reductions in Table 5.18 may be high. If so, the estimates of the cost per kg of phosphorus reduction from farming are low.

While the cost of phosphorus removal from sewage appears to be inexpensive relative to the cost of reducing phosphorus losses from farming, the estimated cost of $12 per kg is equal to approximately $10 per capita annually. The residents of Dryden may not be willing to pay this cost.

The reduction of phosphorus that enters Fall Creek from septic tanks is likely to be a much more difficult task. The scattered population, together with the relatively small number of people to be served would make the cost of sewage collection and treatment relatively high, even though the cost of adding tertiary treatment may be relatively low. In addition, it is questionable whether the sewage plant, even with tertiary treatment would discharge less phosphorus to the stream than currently enters the stream from septic tanks.

SUMMARY

Two questions arise from the economic analysis: (1) What degree of reduction in phosphorus input to Cayuga Lake from Fall Creek and other streams should be achieved to produce the greatest benefits to society? (2) Which source of phosphorus should be reduced first? The results of this study indicate that the answer to the second question is rather clear, at least on the grounds of economic efficiency. Tertiary treatment to remove phosphorus from the effluent of the various sewage treatment plants discharging to the lake or its tributaries appears to be a relatively low cost method, and should be adopted, assuming that reduction of phosphorus input to the lake beyond that achieved by the detergent phosphate ban is needed. If reduction beyond that achieved by the phosphate ban and tertiary treatment is needed, barnyard runoff control appears to be the least costly method of reducing phosphorus loss from farming and should be installed at least for barnyards known to be discharging directly to streams or road ditches. Manure handling practices may be the next least costly method of reducing phosphorus loss from farming. More research is needed relative to the reduction of phosphorus loss than can be achieved by winter storage and spring spreading as well as the increase in manure value likely to be achieved by such a practice. As an alternative to a regulation requiring storage, spreading manure at relatively low rates is likely to keep phosphorus losses at a minimum.

The question of the reduction in phosphorus input to Cayuga Lake necessary to provide the greatest benefit to society cannot be answered at this time. It should be recognized that actions taken to reduce phosphorus losses from farming may produce other results, both positive and negative. For example, reductions of phosphorus related to reducing the acreage of corn or installing conservation practices will be accompanied by reductions in runoff of nitrogen and soil. These reductions may produce benefits to the users of Fall Creek and Cayuga Lake. On the negative side, manure storage undertaken in an effort to reduce runoff losses of phosphorus is likely to increase the odor associated with manure handling, thereby trading a possible improvement of one environmental attribute for a decrease in another. Such environmental trade-offs cannot be ignored.

REFERENCES

Buckman, H. O. and N. C. Brady. 1960. The nature and properties of soils. Macmillan, New York.

Conneman, George J. 1974. Department of Agricultural Economics, Cornell University. Personal communication.

Crowe, Richard. 1974. Soil Conservation Service, Binghampton, N.Y., personal communication.

Edwards, W. M., F. W. Chichester, and L. L. Harrold. 1971. Management of barnlot runoff to improve downstream water quality. Livestock waste management and pollution abatement. The Proceedings of the International Symposium on Livestock Wastes. American Society of Agricultural Engineers, St. Joseph, Michigan.

Environmental Quality Systems, Inc. 1973. Technical and economic review of advanced waste treatment processes, prepared for Office of the Chief of Engineers, U.S. Army Corps of Engineers, Rockville, Maryland.

Federal Register. September 7, 1973. Vol. 38, No. 173.

Hansen, Harold. 1974. Soil Conservation Service, Syracuse, New York, personal communication.

Hutton, Frank Z., Jr. 1971. Soil survey of Cayuga County, New York. United States Department of Agriculture in cooperation with Cornell University Agricultural Experiment Station.

Jacobs, James J. 1972. Economics of water quality management: exemplified by specified pollutants in agricultural runoff. Ph.D Thesis. Iowa State University.

Jacobs, James J. and George L. Casler. 1972. Economic and environmental considerations in dairy manure management systems. A. E. Res. 72–18. Department of Agricultural Economics, Cornell University, Ithaca, N.Y.

Jacobs, James J. and George L. Casler. 1974. Economic and environmental aspects of systems for handling dairy manure. Department of Agricultural Economics, Cornell University, Ithaca, N.Y. (Unpublished manuscript.)

Keup, Lowell E. 1968. Phosphorus in flowing waters. Water Research 2, No. 5: 373–386.

Klausner, S. D. and P. J. Zwerman. 1974. Water quality as influenced by the management of land applied dairy manure. Agronomy Mimeo 74–32. Department of Agronomy, Cornell University, Ithaca, N.Y.

Kumar, Inder Jit and Nicholas L. Clesceri. 1973. Phosphorus removal from waste waters: a cost analysis. Water and Sewage Works. March. pp. 82–91.

Minshall, Neal E., Stanley A. Witzel, and Merle S. Nichols. 1970. Stream enrichment from farm operations. Journal of the Sanitary Engineering Division: Proceedings of the American Society of Civil Engineers. pp. 513–524.

Monti, Randolph P. and Peter T. Silbermann. 1974. Wastewater system alternates: What are they . . . and what cost? Water and Wastes Engineering. May.

Neeley, John A. 1965. Soil survey of Tompkins County, New York. United States Department of Agriculture in cooperation with Cornell University Agricultural Experiment Station.

Seay, Billy D. 1961. Soil survey of Cortland County, New York. United States Department of Agriculture in cooperation with Cornell University Agricultural Experiment Station.

Stanford, G., C. B. England and A. W. Taylor. 1970. Fertilizer use and water quality. U.S. Department of Agriculture, Agriculture Research Service 41–168.

Swader, F. N. 1974. The universal soil loss equation: The key to diagnosing soil erosion. Agronomy Mimeo 74–19, Department of Agronomy, Cornell University, Ithaca, N.Y.

United States Environmental Protection Agency. 1973. Methods for identifying and evaluating the nature and extent of non-point sources of pollutants. EPA-430/9–73–014.

Wischmeier, Walter H., and Dwight D. Smith. 1965. Predicting rainfall-erosion losses from cropland east of the Rocky Mountains: guide for selection of practices for soil and water conservation. Agricultural Research Service. United States Department of Agriculture in cooperation with Purdue University Agricultural Experiment Station. Agriculture Handbook No. 282.

ACKNOWLEDGMENTS

W. Harry Schaffer, Ph.D. student who developed Model D in Chapter 4. Although the model was not directly used in Chapter 5 the authors derived immense benefit from their association with Dr. Schaffer.

Dr. Richard Arnold, Department of Agronomy, Cornell University, for his help in grouping the 131 soil types on the farms surveyed into 19 groups with similar characteristics.

Dr. Douglas Haith, Department of Agricultural Engineering, Cornell University, for his help in computing the cost of phosphorus removal from domestic sewage.

Principal Authors

George L. Casler and James J. Jacobs.

6

Animal Waste Management with Nutrient Control

Chapter 6

ANIMAL WASTE MANAGEMENT
WITH NUTRIENT CONTROL

"All the human and animal manure which the world loses, restored to the land instead of being thrown into the water, would suffice to nourish the world."

Victor Hugo (1802–85)
Les Miserables

INTRODUCTION

This chapter continues the focus on the two major issues of the project: a) conservation and subsequent proper use of nutrients in agriculture, and b) managing nutrients so that any losses have a minimal adverse environmental impact. The efficient utilization of nutrients in animal wastes, especially that of nitrogen, has the potential for assisting crop yields and for savings in commercial fertilizer production.

Feed consumed by farm animals contains about 8 million tons of nitrogen, and the manure produced by these animals contains about 7 million tons of nitrogen. Considerable amounts of the nitrogen excreted by animals are in forms which are quickly converted to ammonia. Volatilization of the ammonia occurs readily as the manure comes into contact with the air. Phosphorus in the manure does not have the same mobility in the environment as nitrogen.

Based upon information presented in Chapter 4, greater than 50% of the nitrogen can be lost from these animal wastes using current waste handling procedures. On a nationwide basis, nitrogen losses from animal wastes may approach 4 million tons annually. This nitrogen has the potential to replace an equivalent amount of fertilizer nitrogen if economically sound means of: a) conserving the ammonia between excretion by the animal and incorporation in the soil, and b) utilizing the recovered nitrogen for crop production were available.

Alternatives to conserve and utilize this nitrogen include:

a) technology to conserve the nitrogen prior to land application,
b) manure application rates that match the ability of crops to utilize the available nitrogen, and
c) manure placed below the soil surface and under conditions that do not lead to denitrification or leaching losses in the soil.

The emphasis of this chapter is on the treatment and stabilization of animal wastes, including nutrient control, prior to their application to land. In the context of this discussion, animal wastes primarily refer to manure but may include spilled feed, water used for cleaning, and other materials used in animal production. A broad spectrum of waste management alternatives is possible.

The intensification of animal production, which was noted for poultry and dairy cattle production in Chapter 1 and elsewhere (Loehr, 1974), has highlighted the need to manage animal wastes to avoid potential pollution problems and has increased the opportunity for greater nutrient control. The larger operations can incorporate technology that may not have been possible or feasible for smaller ones.

In addition the Federal Water Pollution Control Act Amendments of 1972 with the national "zero discharge" goal have elevated the waste management concerns of all production facilities, including that of agriculture, to a higher level. Industrial and agricultural producers are placing waste management decisions at a level comparable to importance granted other production decisions. Individuals having waste management responsibilities are more aware of the technical alternatives and how to properly operate and manage them.

As a result of both the intensification and the heightened concern, waste management approaches that formerly were not considered now are given serious evaluation.

Approaches

The waste management needs at a livestock production facility will be some combination of odor control, waste stabilization, nitrogen control consistent with land disposal constraints, and ease of waste handling and application. Because of reasons given later, the investigations described in this chapter focus on poultry wastes. Many of the fundamentals and approaches that are discussed can be applied to the management of wastes from other animals.

There are many alternatives for satisfactory poultry waste management. There is no one best method for everyone. The possible systems, each having advantages and disadvantages, are noted below.

Deep Pit. The deep pit house has a large basement below the birds, seven to eight feet deep. There is the possibility of ground water seepage into the pits and difficulties in getting enough air circulation to keep the manure dry. Manure tends to be anaerobic and odorous.

High Rise. The house is built above ground with good drainage. Drying of the accumulating manure can be enhanced by circulating fans. A dry material of 40 to 60% moisture can result. Care should be taken to avoid spillage of drinking water for the birds. Odors can be minimal.

Anaerobic. No aeration is provided for these wastes and anaerobic microbial activity results. The wastes generally are handled as a thick slurry or as very moist solids. Severe odor problems are likely.

Aerobic. Aeration equipment entrains oxygen into a liquid containing the wastes. With adequate aeration, odors are nonexistent and the material can be pumped and applied to crop land for both liquid and nutrient use without resulting in complaints due to odors. Examples are the oxidation ditch and aerated tanks or lagoons.

Soil Injection. This is not a treatment system by itself but can be helpful where odors are a problem. The injection units can be used best with anaerobic waste material and have the advantage of immediately incorporating the waste in the soil thus minimizing any volatile losses such as ammonia.

Dehydration. At present there is a limited market in the United States for the dried product. Heat is applied to remove the moisture. At the same time, any volatile compounds also are removed. Odors can be a problem but can be controlled by using afterburners.

Anaerobic Digestion with Methane Production. This involves the controlled digestion of the wastes with capture of the methane. Few such digesters have been in use in the United States. Further technological development is necessary before this process can be applied continuously and effectively at animal production facilities.

Each of these systems has different nutrient conservation efficiencies which are summarized in Table 6.1. The summary refers

Table 6.1. Poultry waste management alternatives—estimated level of nitrogen conservation.

Alternatives	Forms of Nitrogen Present in End-product	Level of Nitrogen Conservation in the Manure	Operational Control[1]
Deep pit	organic, NH_4^+	medium to low	none
High rise	organic, NH_4^+	low	some
Anaerobic	organic, NH_4^+	medium to low	none
Anaerobic with methane fermentation	organic, NH_4^+	high	good
Aerobic	organic, NH_4^+, NO_2^-, NO_3^-	variable, can be high or low depending on operation	good
Soil injection	depends on waste source	high	good
Dehydration	organic	low	none

[1] Control available to the operator to achieve nitrogen conservation.

only to the retention of nitrogen in the manure for each of the alternatives considered. It does not include other losses that might occur when the alternative is used as part of a total waste management system. For example, soil injection by itself conserves most of the nutrients in the waste. However, because injection frequently is used with anaerobic wastes in which prior losses of nitrogen are likely to occur, the overall nitrogen conservation can be medium or low. Additional nitrogen losses can occur after the wastes are injected as a result of leaching and denitrification (Chapter 4).

Aerobic systems have the potential to meet all of the livestock waste management needs identified earlier: odor control, waste stabilization, ease of waste handling, and nitrogen control when needed. These systems have been used with many agricultural and animal wastes and the technology is available. Therefore these studies on engineering methods to control nutrients in animal wastes investigated the feasibility of aerobic systems, specifically the oxidation ditch, for such control.

Nutrient Control

The desirable level of nutrient control is related to the utilization of the nutrients when the wastes ultimately are applied to the land. If the wastes are intended to fertilize crop production, then the objective is to conserve as great a quantity of nutrients as possible. On the other hand, if the applied nutrients are potentially excessive, then it is desirable to reduce the nutrient content of the wastes prior to application.

Chemical precipitation can be used for the removal of phosphorus during treatment of animal wastes (Loehr and Johanson, 1974). However such methods are difficult to manage, appear costly, are not easily integrated into the usual operations of animal production, and result in the transfer of the problem from manure disposal to that of disposal of large volumes of chemical sludge.

Physical and chemical methods of nutrient control, such as chemical precipitation for phosphorus and ammonia desorption or ion exchange for nitrogen can be feasible when a dilute waste is being treated and when the goal is the production of an effluent suitable for discharge to surface waters. Such is not the goal for the treatment and stabilization of animal wastes. The nutrients in animal wastes can be utilized when properly integrated with crop production (Chapters 4 and 5). In addition, federal regulations (Anon, 1974) do not permit the discharge of treated or untreated animal wastes to the nation's waters except where such drainage results from extreme rainfall events.

If there is little likelihood of surface runoff when the wastes are applied, and if the phosphorus has adequate time to react with the soil, phosphorus control through soil conservation practices should be sufficient to protect water quality and provide for utilization of the nutrients. Nitrogen, because of its high solubility, is subject to losses in soil water and runoff (Chapter 3 and 4).

Possible techniques for controlling levels of nitrogen as part of the treatment of animal wastes include ammonia volatilization (desorption) and nitrification followed by denitrification. Ammonia desorption requires a degree of control that is unlikely to be achieved or continuously maintained by those operating animal waste management facilities.

Nitrification-denitrification appears to be a more feasible nitrogen control method. This method can be integrated with other waste management procedures to meet the needs noted earlier. The nitrogen losses from this method will be primarily as nitrogen gas and should not result in subsequent environmental problems. In contrast, the nitrogen lost by desorption will be as ammonia which can be redeposited as a potential contaminant of surface waters.

To understand and use the nitrification-denitrification process, it is necessary to understand the transformations of nitrogen in wastes. Nitrogen in fresh animal excreta primarily is in the organic form: proteins, urea, or uric acid. Microbial waste stabilization systems utilize a sequence of nitrogen transformations, the first step of which is the ammonification of the organic nitrogen.

$$\text{Organic nitrogen} \xrightarrow[\text{(heterotrophs)}]{\text{Ammonification}} \text{NH}_3 \; \underset{\text{(hydrolysis)}}{\overset{\text{H}_2\text{O}}{\rightleftarrows}} \; \text{NH}_4^+ + \text{OH}^-$$

Ammonification of organic nitrogen is accompanied by an increase in pH. If the ammonium concentration and pH are sufficiently high, ammonia volatilization can occur.

Under aerobic conditions, ammonium nitrogen can be microbially oxidized to nitrate by two groups of autotrophic organisms: *Nitrosomonas* and *Nitrobacter*. This process of oxidation of NH_4^+ to NO_3^- is termed nitrification.

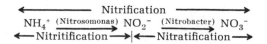

Under anaerobic conditions, nitrite and nitrate can be reduced to nitrogen gas (N_2) by denitrifying organisms. This process is termed denitrification, which may be represented as:

$$\overset{\longleftarrow \quad \text{Denitrification} \quad \longrightarrow}{NO_3^- \longrightarrow NO_2^- \longrightarrow N_2O \longrightarrow N_2}$$

Nitrate   Nitrite   Nitrous
oxide

Although the above sequence implies possible losses of nitrogen as nitrogen oxides, results of studies with animal wastes indicated that losses following denitrification are as nitrogen gas (Loehr et al., 1973). If it were possible to control the balance between nitrification and denitrification, then nitrate could be conserved or converted to nitrogen gas as desired.

Any practical methods for nitrogen control must be economically feasible, convenient to manage, compatible with livestock production operations and must perform satisfactorily under varying environmental conditions in the treatment of concentrated wastes.

ENGINEERING STUDIES

General

The studies reported in this chapter were one component of an extensive series of investigations on engineering methods to control nutrients in animal wastes. The total investigations spanned over six years and were supported by the Environmental Protection Agency and its predecessors, by the College of Agriculture and Life Sciences and by the Rockefeller Foundation. Prior research was basic studies aimed at identifying the feasible technical alternatives. They included laboratory investigations on phosphorus control by chemical precipitation, on ammonia desorption in aeration systems, and on nitrification and denitrification.

The results of these initial studies have been published (Prakasam et al., 1974; Loehr et al., 1975; Srinath and Loehr, 1974) and were used as the basis for the type of experimentation conducted as part of this project. Throughout the period of these investigations, complementary nutrient control studies supported by other organizations continued. The personnel from all of these engineering studies discussed activities and results. As a result, these and other studies were mutually supportive and catalytic such that all progressed at a more rapid pace than otherwise would have occurred.

Detailed studies at Cornell on the performance of oxidation

ditches for treating poultry wastes, and studies elsewhere on treating other livestock wastes, suggested that an oxidation ditch would meet the livestock waste management needs identified earlier. This system is a viable method of biologically treating agricultural wastes with several advantages:

(1) odors may be eliminated or controlled;
(2) labor needs are modest;
(3) biodegradation of the wastes can be controlled;
(4) installation and operation of the process are compatible with intensive animal production units;
(5) efficiently operating units are not limited to particular geographical areas;
(6) the system is applicable to all animal wastes, although its major use has been with small animals.

Consequently, a major objective of this study was to explore the use of oxidation ditches as a means for controlling the levels of nitrogen in the waste ultimately applied to the land.

Poultry wastes were studied as part of this work because of the excellent available facilities and the ease of managing small animals.

Objectives of This Study

Although the biological nitrification-denitrification process has been utilized in the treatment of municipal and industrial wastes, it has not been generally applied in agriculture. The studies described in this chapter evaluated the viability of using this process in treating animal wastes.

The specific objectives of the research were to concurrently (1) identify factors affecting nitrogen losses from oxidation ditches; (2) examine factors influencing nitrification; (3) assess the performance of systems in the field. This research was undertaken to extend the information that had been or was being obtained by the other complementary studies.

In order to use the available time as effectively as possible, a number of simultaneous studies were undertaken. The above objectives were achieved by parallel investigations at the laboratory scale, pilot plant, and field levels. The laboratory experiments were conducted to gain understanding of the biological processes involved, and in particular, of the fundamental characteristics of the processes occurring in the aerobic stabilization of animal wastes. This information assisted the evaluation of waste treatment systems for either maximum conservation or maximum removal of nitrogen. The objectives of the pilot plant studies and observations on full-scale systems (field units) were to: (1) further verify the concepts; (2) examine problems associated with large-scale operations; and (3) evaluate full-scale performance. The field units were not designed to achieve a specified degree of nitrogen control. The nitrogen control that was achieved by these operating units was assessed.

In the bench-scale investigation, and to a degree in the pilot plant units, it was possible to have relatively tight control of different conditions. In actual production units, this control is not possible. The reality of actual operating conditions must be incorporated in the design and application of the feasible technologies, and this was done through the evaluation of the field units. The investigations of the field units served as a check on concepts developed from laboratory and pilot plant results and indicated some of the managerial problems, such as maintenance, operation, and monitoring of the system.

Facilities and General Methods

The pilot plant studies were conducted at the Agricultural Waste Management Laboratory on the Cornell University Campus. The oxidation ditch at the pilot plant received wastes from approximately 250 birds each day. The field unit studies were conducted in the following poultry operations: (1) a commercial egg production operation near Ithaca, owned by Mr. Charles Houghton; and (2) a commercial egg production operation at Camillus, New York, owned by Manorcrest Farms.

The applicability of these poultry waste stabilization systems to the management of wastes produced by other animals was demonstrated at the United States Department of Agriculture experimental fur animal production facility located at Ithaca, New York.

Throughout the investigations, the physical, chemical, and biochemical characteristics of both waste and treated effluent were monitored. Because of the distances between the field installations and the laboratory facilities, sampling was limited to twice per week at the production units. The analytical procedures employed were those evaluated by project personnel and found suitable. The parameters measured and procedures used are identified in Table 6.2. Solids degradation, COD or BOD reduction and nitrogen transformations were derived from the data.

Table 6.2. Parameters measured in these studies.

Parameter	Analytical Method
total, suspended and volatile solids	Standard Methods, 1971
COD	Jeris, 1967
BOD	Standard Methods, 1971
all forms of nitrogen[1]	Prakasam et al., 1972
dissolved oxygen	dissolved oxygen probe
temperature	thermometer
pH	pH meter

[1] Organic, NH_4^+, NO_2^-, and NO_3^- nitrogen.

STUDIES ON NITRIFICATION REACTIONS

The objective of these studies was to identify the influence various factors had on the rate and extent of nitrification. Such studies were important since an understanding of nitrification is critical to the control of nitrogen by nitrification-denitrification processes.

Nitrification

The nitrogen transformations resulting in nitrification have been identified earlier. Certain species of heterotrophs can cause nitrification, but were not of practical significance in these studies. The nitrification which occurs in aerobic treatment systems is due to autotrophic nitrifying organisms. Seven genera of autotrophic nitrifying microorganisms are known, but bacteria belonging to the genera *Nitrosomonas* and *Nitrobacter* are those considered to be the predominant nitrifying organisms in nature.

The rates of ammonification and nitrification are influenced by factors that include: (1) concentration of dissolved oxygen, (2) pH, (3) population of appropriate microorganisms, and (4) temperature. Several compounds inhibit nitrification—mercaptans, nitrous acid, and undissociated ammonia are among those that can be encountered in waste treatment systems. It was found in other work that the inhibitory effects of nitrous acid and undissociated ammonia could be controlled by varying operating conditions (Anthonisen, 1974).

Experimental Procedures

The above variables were closely controlled and monitored. The dissolved oxygen concentration in the reactors was monitored routinely to assure that aerobic conditions existed and that nitrification was not inhibited by a lack of adequate aeration. The dissolved oxygen levels were always above 0.5 mg/l and generally were above 1.5 mg/l. The substrate concentrations in the reactors were varied to determine effects this variable might have on the reactions and to avoid any possible inhibition due to lack of substrate at high reaction rates.

All experiments were of the batch type. The substrates employed, the ranges of their concentration and other variables examined are noted in Table 6.3. The experimental approach was to examine the nitrification steps individually and independently. The important parameters that were varied and controlled were: (1) pH, (2) temperature, and (3) concentration of active microorganisms in nitrifying systems. A major purpose of the experiments was to identify not only the effect of varying these parameters but also to identify conditions that would limit the process.

Nitrifying Organisms

An enrichment culture of nitrifying organisms was prepared by inoculating the culture medium (Table 6.4) with the mixed liquor from an oxidation ditch in which nitrification was taking place. The purpose of the enrichment culture was to assure an adequate population of nitrifying organisms in each experiment. In these and subsequent studies, no attempt was made to isolate and identify specific types of bacteria. In the remainder of this chapter, the ammonium- and nitrite-oxidizing organisms will be termed respectively as nitrosomonads and nitrobacter, with the understanding that the intent is to indicate the type and not the specific genera of the organisms.

Ammonium sulfate and sodium nitrite were used separately at concentrations of 1000 mg N/l as substrates for these cultures. The inoculated culture vessels were aerated and tested daily for residual substrate. When the substrate was nearly all utilized, aeration was stopped, microbial mass allowed to settle, and the supernatant medium decanted. Fresh medium was added and aeration continued. By repeating this "fill and draw" procedure of the medium, reserve cultures of nitrosomonad and nitrobacter were prepared. The pH of the nitrosomonad culture was adjusted to initial 7.6–7.8 levels. Since NO_2^- oxidation caused no change in pH, adjustment of pH was not required for nitrobacter cultures.

In preparation for specific experiments, the stock cultures were

Table 6.3. Ranges of variables examined in the studies on kinetics of ammonification, nitritification and nitratification.

Variable	Ammonification	Nitritification	Nitratification
Substrate concentration (mgN/l)	urea & casein (10:1) = 500–1600 uric acid & casein (10:1) = 700–1000 poultry waste = 800–1200	$(NH_4)_2SO_4$ = 100–1000	$NaNO_2$ = 100–1000
Temperature (°C)	20	9–34	9–34
pH	6.5–9.0	6.0–8.5	6.0–8.5
Suspended volatile solids (SVS, mg/l)	1000–8600	250–4000	250–3000

Table 6.4. Basic culture medium (g/l).

Sodium dibasic phosphate (hepta hydrate)	2.25
Sodium chloride (anhydrous)	6.75
Potassium dibasic phosphate (anhydrous)	5.00
Magnesium sulfate (hepta hydrate)	0.15
Ferric chloride	7.5 mg/l
Ammonium molybdate solution[1]	0.25 ml/l
Ammonium sulfate or sodium nitrite	as desired
Water	Cornell hot tap water
pH of medium	7.6–7.8

[1] 25 g $(NH_4)_6MO_7O_{24} \cdot 4H_2O$ per liter.

kept in suspension in fresh medium adjusted to desired pH conditions, and were allowed three weeks to adapt to the new environment before specific tests were conducted. In adapting cultures to different temperatures, they were brought to the desired temperature by small changes, allowing the organisms to adjust slowly. Complete adjustment to environmental conditions was assumed when consecutive tests showed reasonable duplication in oxidation rates. Once acclimated to the test conditions, the organisms were used repeatedly by settling out microbial mass, discarding the spent medium and resuspending the settled mass in fresh medium.

In each test the following were monitored: (1) substrate and reaction product concentration to establish the rate of the reaction, (2) dissolved oxygen concentration, (3) pH, (4) temperature, and (5) microbial mass. The concentration of mixed liquor volatile suspended solids (MLVSS) was used as a measure of microbial mass. To increase or decrease the concentration of microbial mass to a desired level, either the volume of substrate was varied or some of the microorganisms were wasted.

The pH was adjusted to desired levels by addition of either concentrated sulfuric acid or a saturated solution of sodium carbonate and was maintained at those levels by an automatic control system using suitable acid and base solutions.

The temperature was controlled to desired levels using a walk-in incubator and temperature-controlled water baths.

General Results of Nitrification Studies

In experiments in which only temperature was controlled, the rate of disappearance of organic nitrogen appeared proportional to the amount present at any given time (Figure 6.1). The NH_4^+ concentration increased initially, with subsequent reduction in its concentration coinciding with the increase in the rate of formation of oxidized nitrogen ($NO_2^- + NO_3^-$). The pH of the system increased with the initial increase in NH_4^+ concentration and decreased when nitrification rates increased.

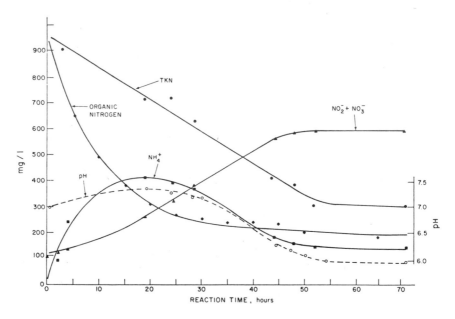

Figure 6.1. Typical reaction profile with oxidation ditch liquor—batch experiments.

In initial experiments at controlled, low pH values, substrate concentration (NH_4^+ or NO_2^-) appeared to inhibit the reaction rate. In all experiments conducted at pH levels greater than 6.5, the initial substrate concentration was 1000 mg N/l. Where inhibition appeared possible, lower substrate concentrations were used to avoid this effect. There appeared to be inhibition when the concentration of nitrous acid (HNO_2) in the system was about 2.6 mg HNO_2-N/l.

Ammonia Oxidation

When pH, temperature, and microbial mass in the systems were held constant, the NH_4^+ oxidation rate was independent of the NH_4^+ concentration (Figure 6.2):

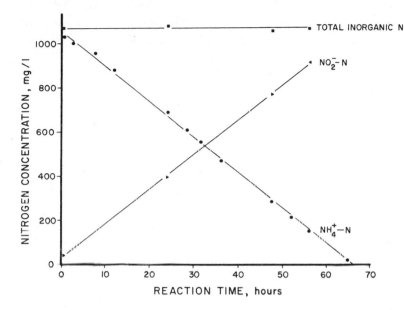

Figure 6.2. Typical NH_4^+-N oxidation changes at a controlled pH and temperature.

$$\frac{dA}{dt} = -k_2^* \tag{1}$$

where,

A = the NH_4^+-N concentration
k_2^* = the reaction rate
t = the reaction time

Effect of Microbial Concentration

The effect of microbial concentration on the NH_4^+ oxidation reaction is illustrated in Figure 6.3. Each point represents an experiment in which the rate of NH_4^+-N oxidation was determined at a noted microbial concentration. The oxidation rate, k_2^*, is the rate of NH_4^+-N depletion as noted in Equation 1 and Figure 6.2.

The data in Figure 6.3 can be mathematically represented as:

$$[1/k_2^*] = [1/k_2^*{}_{max}] + [1/(k_2 \cdot Sa_m)] \tag{2}$$

where,

k_2 = NH_4 oxidation rate constant (per hour)
k_2^* = NH_4 oxidation rate, mg/l/hr
$k_2^*{}_{max}$ = maximum NH_4 oxidation rate, mg/l/hr
Sa_m = concentration of nitrosomonads (SVS), mg/l

The data in Figure 6.3 were analyzed by plotting the reciprocal of the oxidation rate against the reciprocal of the microbial concentration (Figure 6.4). A linear relationship between them was obtained. For the five levels investigated, ranging from 6.0 to 8.0, the correlation coefficient in all cases was greater than or equal to 0.92. Thus the equation noted in Figure 6.3 is a reasonable representation of the data.

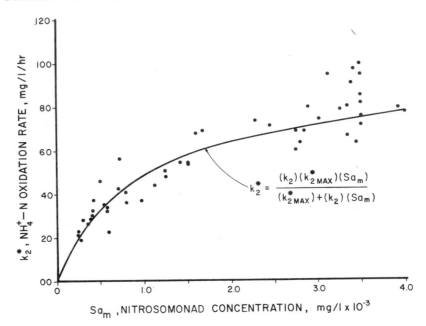

$$k_2^* = \frac{(k_2)(k_{2\,MAX}^*)(Sa_m)}{(k_{2\,MAX}^*)+(k_2)(Sa_m)}$$

Figure 6.3. Typical ammonium nitrogen oxidation changes at varying microbial concentration, controlled pH and temperature.

Effect of pH

The ammonia oxidation rate is a function of (1) a maximum oxidation rate $k_{2\,max}^*$; (2) a reaction rate constant, k_2; and (3) the microbial concentration, Sa_m. The effect of pH on $k_{2\,max}^*$ and k_2 is noted in Figure 6.5. The maximum oxidation, 114 mg N/l/hr, was obtained at a pH of 7.5.

The microbial concentration will affect the maximum oxidation rate (Figure 6.6), but the optimum pH appeared somewhat independent of microbial concentration.

The general effect of pH on the ammonia oxidation is to decrease

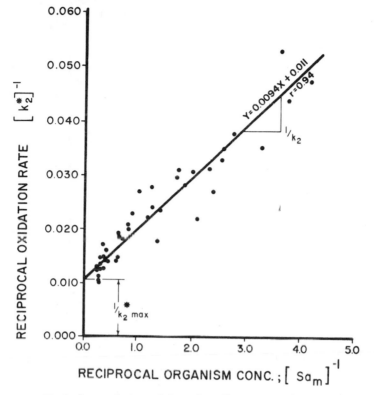

Figure 6.4. Typical correlation of data describing ammonium-nitrogen oxidation.

the oxidation rate as the pH decreases. This occurrence explains the exponential nature of the ammonia changes where pH is uncontrolled.

Effect of Temperature

The semi-log plot of reaction rate constant (k_2) against the reciprocal of temperature (degrees Kelvin) yielded a straight line (Figure 6.7) such that:

$$k_2 = [k_2]_0 \cdot \text{Exp} [-E/RT] \tag{3}$$

The values for the energy (E) were calculated from Figure 6.7. The activation energy appeared lowest about pH 7.5, which may account for the high rate of oxidation at that pH. The reaction rate decreased as the temperature decreased.

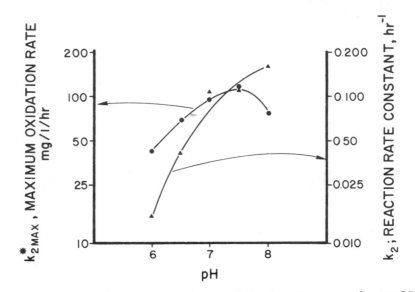

Figure 6.5. Effect of pH on ammonium oxidation by nitrosomonads at 20°C. Microbial mass held constant.

Nitrite Oxidation

General

The factors that affected ammonia oxidation also were found to affect nitrite oxidation, *i.e.*, pH, temperature, microbial concentration, and presence of inhibitory substances. Analysis of the nitrite oxidation results indicated that the data could be analyzed in a manner similar to that used in analyzing the information on ammonia oxidation.

With temperature, pH, and microbial mass kept constant, the nitrite oxidation rate was independent of the NO_2^- concentration (Figure 6.8):

$$\frac{dm}{dt} = -k_3^* \tag{4}$$

where,

  m = concentration of nitrite (mg N/l)
  t = reaction time (hr)
  k_3^* = reaction rate constant, and is the slope of the nitrite depletion curve (mg/l/hr)

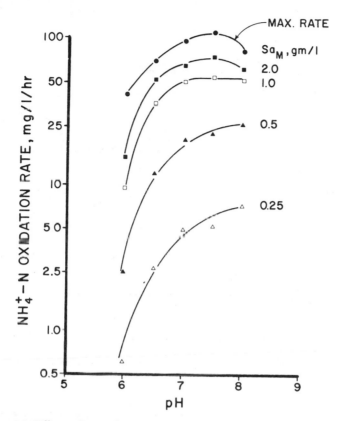

Figure 6.6. Effect of microbial concentration on ammonium oxidation rate at 20°C and over a pH range of 6 to 9.

Effect of Nitrobacter Concentration

At a given pH and temperature, the value of k_3^* was dependent on the concentration of nitrobacter and Equation 5 was found to provide a reasonable estimate of the data (Figure 6.9).

$$k_3^* = \frac{k_3^*{}_{max} \cdot k_3 Sa_B}{k_3^*{}_{max} + k_3 Sa_B} \qquad (5)$$

or

$$[1/k_3^*] = [1/k_3^*{}_{max}] + [1/(k_3 \cdot Sa_B)] \qquad (6)$$

where,

$k_3^*{}_{max}$ = maximum NO_2 oxidation rate, mg N/l/hr

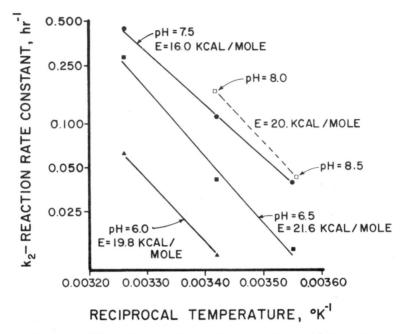

Figure 6.7. Effect of temperature on the ammonium oxidation rate.

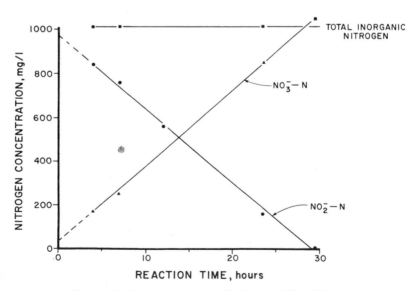

Figure 6.8. Typical nitrite oxidation at 20°C, pH 7.3.

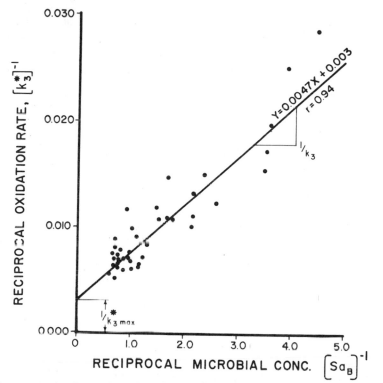

Figure 6.9. Analysis of results of nitratification study, nitrite oxidation, pH 7.3, temperature 20°C.

k_3 = NO_2 reaction rate constant (per hr)
k_3^* = NO_2 oxidation rate, mg N/l/hr
Sa_B = concentration of nitrobacter (SVS), mg/l

Effects of pH and Temperature

The experimentally observed values of k_3 and $k_3^*{}_{max}$ are summarized in Table 6.5. The value of k_3^* is influenced by pH and that of k_3 by both pH and temperature. The effects of pH on nitrite oxidation were determined by collectively examining the data.

The k_3–pH relationship has a maximum in the pH range of 7.0 to 7.5 (Figure 6.10). At 20°C high rates of oxidation could be expected at pH values as high as 8.5.

The most significant feature of the temperature relationships (Figure 6.11) is that the lowest value of the activation energy is at pH 7.3. Significantly higher values occurred at pH values of 6.5 and 8.5 indicating a possible reason for the lower rates of oxidation at these values.

Table 6.5. Nitrite oxidation rates at different temperatures and pH values.

Temperature °C	pH	$k_3{}^*{}_{max}$ (mg N/l/hr)	k_3 (per hr)
9	8.5	358	0.013
15	7.3	278	0.193
	6.5	238	0.052
20	8.5	348	0.212
	8.0	388	0.108
	7.3	250	0.292
	6.5	238	0.079
	6.0	219	0.032
27	8.5	379	0.222
	7.3	230	0.281
34	7.5	110	0.146
	7.3	330	0.378
	7.0	225	0.416
	6.5	125	0.057

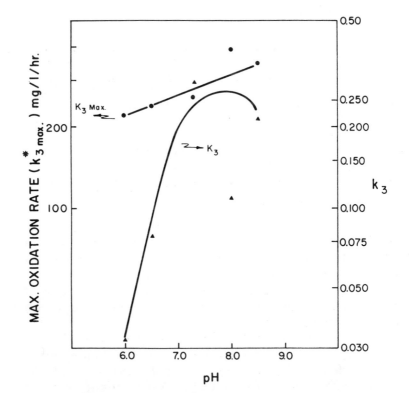

Figure 6.10. Effect of pH on nitrite oxidation at 20°C.

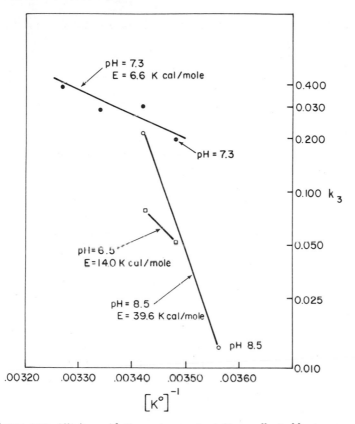

Figure 6.11. Nitrite oxidation rate constant, K_3 as affected by temperature.

In some experiments at low pH values, substrate concentration appeared to inhibit the reaction rate. In all experiments conducted at pH levels of 6.5 and 6.0, the substrate concentrations were about 450 and 200 mg N/l, respectively. These lower concentrations were used to avoid possible inhibition by undissociated nitrous acid.

Relevance of Laboratory Nitrification Studies

An adequate understanding of nitrification is critical to nitrogen control by nitrification-denitrification since if nitrification does not take place, nitrogen loss by denitrification will not result. The information on the basic nitrification reactions provided valuable insight to the nitrification process and the factors that affect its performance. The rates of reaction under different pH, temperature, and substrate conditions, and with different concentrations of nitrifying bacteria were identified. Potentially inhibitory conditions were observed and noted.

In this way, the nitrification results that do occur in the "real world" can be understood in terms of the fundamentals of the reactions rather than considering the system as an unknown "black box." Examples may assist in understanding the use of this information. Less than optimum pH or temperature conditions will result in lower microbial reaction rates. However, this can be compensated for by increasing the total reaction time (solids retention time) in the biological treatment process. The desired nitrification efficiency can be achieved by increasing the time that the nitrifying organisms have to oxidize the wastes. One can also increase the quantity of microorganisms in the treatment process so that although the microbial rate is less than optimum, the larger numbers of nitrifying organisms will achieve the desired efficiency.

Should the treatment system not be obtaining the desired effluent quality, the performances can be evaluated in terms of the fundamental factors affecting the system and remedial measures attempted. Inhibition conditions can be recognized and possibly relieved. Microbial reaction times can be increased. Process loading rates can be decreased.

The relevance of these results lies in permitting those designing and operating nitrification processes to better understand the process and assure its successful application.

STUDIES ON ACTIVE MASS
OF NITRIFYING ORGANISMS

The design and assessment of the performance of nitrification processes requires knowledge of the active mass of nitrifying organisms in the processes. Mixed liquor volatile suspended solids (MLVSS) is the most common method of estimating the concentration of active organisms in biological treatment systems. This is not, however, a useful measure of the active nitrifying mass.

MLVSS is a measure of the particulate matter in a treatment process and can include inert volatile matter and unmetabolized wastewater solids as well as microorganisms. A more specific measure of the active mass of nitrifying organisms should be used for the design and operation of nitrification systems. Microbiological methods exist that can be used to estimate the numbers of nitrifying organisms, but they are not practical for the design or operation of nitrifying systems.

As part of this research, a method was developed to estimate the nitrifying active mass. The method appears to have broad application for the design of nitrification processes and in the assessment of their performance.

Method

The microbial oxidation of ammonium and nitrite is independent of the concentration of each compound as noted in the previous section. The concentration of the reaction product (nitrite in the case of ammonium oxidation by nitrosomonads, and nitrate in the case of nitrite oxidation by nitrobacter) at any time (t) can be expressed as:

$$C_t = C_o + kt \tag{7}$$

where C_t and C_o are the concentrations (expressed as mg/l) of the reaction product at times t and zero, respectively; and k is the oxidation rate constant. The units of k are mg per liter per unit time.

When dissolved oxygen concentrations above the minimal level required to sustain nitrification are maintained, the rate of reaction at a given temperature and pH is directly dependent upon the concentration of the active organisms. Under these conditions, k is dependent on the concentration of nitrifiers in the system and can be used as part of the estimate of the nitrifying mass.

The total nitrogen content of the nitrifying organisms was used as part of the estimate of the nitrifying active mass. In cultures of nitrifying organisms growing at a logarithmic rate, the extraneous sources of organic nitrogen are negligible. Therefore, it was assumed that the cellular total nitrogen content of such cultures is a reasonable estimate of the quantity of bacterial cells. The microbial solids were washed in a saline solution before the total Kjeldahl nitrogen was measured.

The specific activity of substrate oxidation, i.e., k per concentration of nitrifying cell mass (mg/l/hr/mg TKN/l), should be constant at any given temperature and pH. Therefore, the pH-specific activity relationship at any standard temperature is a useful tool to compare nitrification rates and thus the nitrifying activity of treatment processes.

Nitrifying systems under field conditions have mixed microbial populations. In such systems, a TKN analysis would measure the TKN of any wastes plus that of the heterotrophs and autotrophs. Therefore, specific activity values in an actual treatment system would be lower than those obtained with pure cultures of nitrifying organisms under the same conditions. The ratio between specific activities in a treatment system and a pure culture system, under comparable experimental conditions, could be interpreted as the fraction of total microbial TKN (active mass) that is contributed by the autotrophic nitrifiers, *i.e.*,

$$
\begin{array}{c}
\text{Fraction of} \\
\text{Nitrifying Active Mass}
\end{array}
=
\dfrac{\begin{array}{c}\text{specific activity of the MLVSS} \\ \text{in a treatment process}\end{array}}{\begin{array}{c}\text{specific activity of a standard} \\ \text{nitrifying pure culture}\end{array}}
\tag{8}
$$

Implicit to this approach is the assumption that the total nitrogen content of all microbial cells expressed on a dry weight basis is comparable and that the nitrifiers in the mixed cultures exhibit reaction rates similar to those in pure cultures operating under similar conditions.

This fraction (Equation 8) was termed the specific activity coefficient of the nitrifiers in the treatment system. The product of the specific activity coefficient and total Kjeldahl nitrogen of the treatment system was used as an estimate of the nitrifying active mass. The relationship between the specific activity coefficient, nitrifying microbial mass, and total Kjeldahl nitrogen can be expressed as:

$$
\begin{aligned}
\text{Specific Activity Coefficient} &= \dfrac{\text{specific activity of test sample}}{\text{specific activity of standard pure culture}} \\[2mm]
&= \dfrac{\text{mg N oxidized per hour per liter/mg TKN per liter (in test sample)}}{\text{mg N oxidized per hour per liter/mg TKN per liter (in pure culture)}}
\end{aligned}
\tag{9}
$$

$$
\begin{array}{c}
\text{Nitrifying Active Mass} \\
\text{of the Sample (mg TKN/l)}
\end{array}
=
\left[\begin{array}{c}\text{specific activity} \\ \text{coefficient}\end{array}\right]
\times
\left[\begin{array}{c}\text{concentration of TKN of} \\ \text{cells in test sample (mg/l)}\end{array}\right]
\tag{10}
$$

This approach was evaluated under laboratory conditions and utilized with actual treatment systems to determine its feasibility. The microorganisms used were cultures of *Nitrosomonas europaea, Nitrobacter agilis* and *Nitrobacter winogradskyi* obtained from the type culture collection of the Oceanography Institute, Woods Hole, Massachusetts.

All the experiments were conducted at room temperature (20–21°C). The effect of microbial concentration, pH, and substrate concentration on the reaction rates was examined. The ranges of the tested parameters are given in Table 6.6.

Table 6.6. Parameters evaluated in the nitrifying active mass experiments.

		Range of Parameter Examined	
Organism	pH	Substrate Concentration (mg N/l)	Cell Mass in the System (mg TKN/l)
N. europaea	6.0–9.0	50–1000	5–60
N. agilis	6.0–9.0	100–1500	5–25
N. winogradskyi	6.0–9.0	100–1500	5–25
Oxidation ditch Mixed liquor Solids	6.0–9.0	10–500	30–120

Results

The total Kjeldahl nitrogen and volatile solids content of the harvested and washed cells in the logarithmic phase of growth were determined at random during the study. The volatile solids to total Kjeldahl nitrogen ratio for the *Nitrosomonas* and *Nitrobacter* cultures was 13:1 and 15:1, respectively.

Within the concentration ranges of 50 to 1000 mg NH_4^+–N/l and 100 to 1500 mg NO_2^-–N/l tested in this study, the oxidation of the substrates was independent of substrate concentration and substantiated the results reported in an earlier section. Under comparable conditions, the rates obtained with the oxidation ditch mixed liquor were significantly lower than the rates obtained in pure culture systems (Table 6.7).

The effect of pH on the oxidation of NH_4^+ and NO_2^- was evaluated using the stock cultures (Figure 6.12). Oxidation rates for both *Nitrosomonas* and *Nitrobacter* were maximum in the pH range of 7.4 to 7.9. This information again substantiated the results presented earlier. The specific activity using oxidation ditch mixed liquor at different pH values (Figure 6.13) was the same as those obtained with the stock cultures. The optimum activity of the oxidation ditch mixed liquor solids persisted over a wider range of pH than those exhibited by pure cultures. For purposes of comparing the nitrifying activity of reference cultures of nitrifiers, a pH in the range of 7.4 to 7.9 should be used since this is the optimum range identified for both mixed liquor and stock cultures.

Based upon these results, a procedure estimating the quantity of nitrifying active mass in a mixed microbial treatment system was developed. The details of the procedure are noted in the Appendix.

Table 6.7. Oxidation of ammonia and nitrite.

Cell Nitrogen Content in the System (mg N/l)	Rate of Oxidation of Substrate (mg N/l/hr)			Specific Activity (mg N oxidized per mg cell N per hr)
	pH	NH_4^+	NO_2^-	
N. europaea				
58.0	7.8	36.8	—	0.64
38.4	8.4	10.4	—	0.27
5.2	6.5	0.5	—	0.01
5.2	6.8	1.2	—	0.22
5.2	7.2	2.7	—	0.51
N. agilis				
7.0	7.8	—	12.9	1.8
13	8.0	—	22.1	1.7
19	8.2	—	29.9	1.6
N. winogradskyi				
2.3	7.8	—	4.8	2.1
7.0	8.1	—	12.0	1.7
4.4	6.8	—	4.8	1.1
Oxidation ditch solids				
70	7.8	0.26	—	3.7×10^{-3}
70	7.8	—	0.36	5.1×10^{-3}
39	7.8	0.027	—	3.5×10^{-4}
79	7.8	0.019	—	4.7×10^{-4}
39	7.8	—	3.95	5.0×10^{-2}
79	7.8	—	2.40	6.1×10^{-2}
110	7.0	0.25	—	2.3×10^{-3}
90	7.4	0.53	—	5.8×10^{-3}
100	7.8	0.74	—	7.4×10^{-3}
100	8.0	0.42	—	4.2×10^{-3}

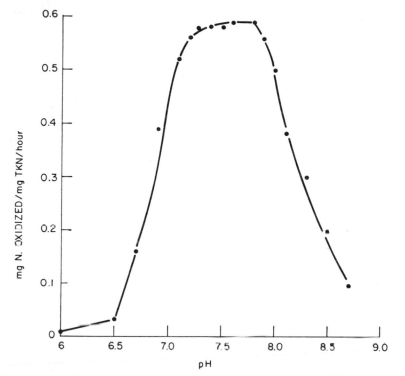

Figure 6.12. Effect of pH on the specific nitrification activity of *Nitrosomonas*.

The procedure is in use in our laboratories as part of additional laboratory and field-scale nitrification studies.

The procedure was used in this study as time permitted. An experiment was conducted to find the effect of anoxic (zero dissolved oxygen) conditions on the nitrifying activity of the oxidation ditch mixed liquor. When the samples of oxidation ditch liquor were held under anoxic conditions for over two months, the nitrifying activity decreased in a nonlinear pattern (Table 6.8). Although the activity

Table 6.8. Nitrifying activity of oxidation ditch liquor under anoxic conditions.

Days	Nitrifying Activity (mg NH_4^+ oxidized per mg TKN)
0	48×10^{-4}
2	26×10^{-4}
9	5×10^{-4}
25	0.4×10^{-4}
50	0.2×10^{-4}

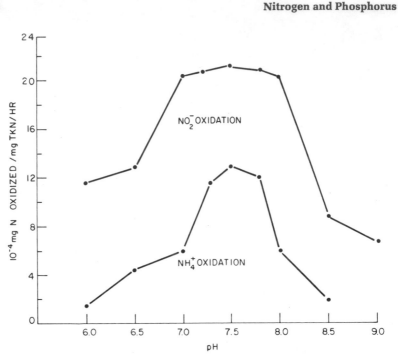

Figure 6.13. Substrate oxidation by a nitratifying oxidation ditch liquor at different pH values.

of nitrite-forming organisms decreased, active organisms persisted despite these anoxic conditions.

These results suggest the nitrifying bacteria are sufficiently robust to withstand periods of oxygen deficiency such as would occur in a nitrification-denitrification system described in subsequent portions of this chapter.

Summary of Nitrifying Mass Studies

The nitrifying active mass procedure provides a better estimate of the quantity of active organisms in a nitrifying system than MLVSS. The procedures can be used to understand how nitrifying organisms perform under controlled environmental and operating conditions. It will also permit the development of rational design and operating criteria based on the amount of nitrifying organisms rather than on the amount of volatile solids.

The results reported in this and earlier sections provide an understanding of the factors that are necessary for the proper design, operation and monitoring of full-scale nitrification systems. In the following sections, these factors added to our understanding of waste treatment systems that utilize the nitrification-denitrification process.

PILOT-SCALE STUDIES OF NITROGEN CONTROL IN OXIDATION DITCHES

Nitrogen Losses

To summarize material presented earlier, losses of nitrogen from a treatment system can occur by volatilization of ammonia and denitrification of oxidized nitrogen. Volatilization of ammonia takes place when the pH of the system is alkaline and undissociated ammonia is present. Loss of nitrogen by denitrification occurs under anoxic conditions during which nitrite or nitrate are used as hydrogen acceptors by the facultative heterotrophs. Nitrification reduces the pH which is unfavorable for nitrogen losses due to ammonia volatilization.

Results of a closely related research study conducted by participants in this work (Prakasam et al., 1974; Loehr et al., 1975) have shown that nitrogen losses occur in nitrifying systems that contain measurable dissolved oxygen concentrations. Some of the factors governing such losses have been examined. By maintaining a minimum dissolved oxygen concentration in the mixed liquor, and manipulating loading rate, nitrogen losses by denitrification may be increased while maintaining an actively nitrifying system.

Modes of Operation of Oxidation Ditch

Attempts were made to develop procedures for operating an oxidation ditch to achieve varying degrees of nitrogen removal. The following modes of operation were studied (Figure 6.14) using the pilot plant oxidation ditch:

I. Continuous rotor operation and no intentional wasting of effluent, i.e., as an aerated holding tank with continuous addition of solids;

II. Maintenance of a solids equilibrium condition and subjecting the mixed liquor to intermittent denitrification.

III. Incorporation of a settling tank achieving denitrification of the recycled effluent.

IV. Curtailed periods of rotor aeration. The rotor was connected via a switch which controlled the time of rotor operation.

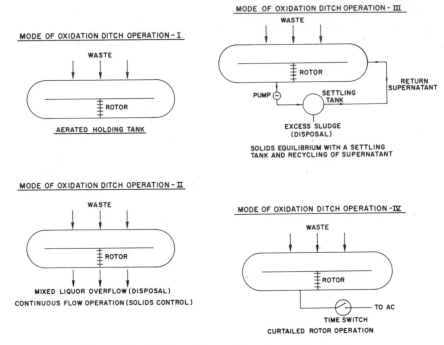

Figure 6.14. Modes of operation of oxidation ditch.

Calculation of Nitrogen Loss

The loss of nitrogen from the oxidation ditch in a given mode of operation was computed using a mass balance. The loss was the difference between the amount of nitrogen that entered the system and the nitrogen present in the mixed liquor plus any nitrogen that was removed deliberately such as that wasted via the sludge from a settling tank. The calculated losses were mostly due to denitrification, since the oxidation ditch was operated as a nitrifying system.

Results

Initial losses of nitrogen were due to ammonia volatilization. These became minimal once nitrification was established and the pH dropped below 7.0. Except for an initial ammonia odor, the oxidation ditch performed without odor during the various modes of operation. Some foaming occurred during transition from one mode of operation to another, but was not a problem since it quickly subsided.

Operational Mode I

In this mode (Table 6.9) the system was operated as an aerated holding tank with continuous input of manure from the birds. No

Table 6.9. Operational parameters of oxidation ditch—Mode of Operation I.

Number of birds	227
Number of days operated	276
Immersion depth of rotor	6 in. (15.2 cm)
Liquid volume	2000 gallons ($\simeq$7600 liters)
Liquid depth	20 in. (51 cm)

odorous conditions were noted and nitrites rather than nitrates predominated during the entire period. Inhibitory concentrations of free ammonia occurred in the mixed liquor and was thought to be the reason for the incomplete nitrification. The solids accumulated to a concentration of 8.2% at which time the rotor was unable to move them. Nitrification was sustained at all solids concentrations. Material balances of nitrogen and COD are presented in Table 6.10.

Table 6.10. Nitrogen and COD balance—Mode of Operation I.

	Nitrogen, kg	COD, kg
Input	51.7	1222
Accounted	35.6	459
Unaccounted or losses	16.1	763
Loss (%)	31.1	62.5

In spite of the high dissolved oxygen concentration in the mixed liquor (>5 mg/l) and active nitrification, about 30% of the input nitrogen was lost. This was attributed to denitrification occurring in the mixed liquor. The COD loss (removal) indicated that the waste was well stabilized.

Operational Mode II

The ditch was operated as a continuous-flow unit with intermittent rotor stoppages to produce denitrification. The concentration of solids was maintained at a constant level by deliberately wasting mixed liquor.

Two *in situ* denitrification studies were made by stopping the rotor. In the first study, continuous-flow operation was temporarily suspended and no overflow was permitted. The birds continued to add wastes to the ditch, thus adding an additional oxygen demand. During the second study, some overflow was permitted. In both studies the rotor was turned on briefly each day to mix the ditch contents. The total nitrogen losses in the different phases are shown in Figure 6.15.

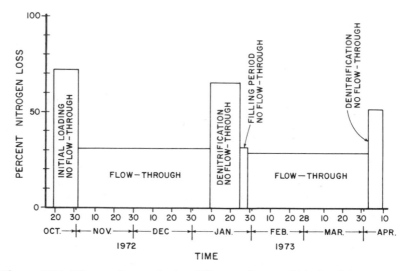

Figure 6.15. Nitrogen losses during different phases of Mode of Operation II.

During the "flow-through" stages of the oxidation ditch, nitrification occurred. At the same time, significant amounts of nitrogen, again about 30%, were lost. This was assumed due to denitrification in anoxic zones of the unit since ammonia volatilization was negligible. This suggested that nitrogen losses may be increased if localized anoxic conditions for denitrification are created. Addition of more raw manure will increase both the solids content and the oxygen demand of the system and increase the probability for anoxic conditions in the system.

Operational Mode III

In this mode, attempts were made to control the solids content of the mixed liquor and thus provide for control of the anoxic conditions. The mixed liquor from the oxidation ditch was pumped intermittently into a settling tank with the supernatant liquid returned to the ditch. The sludge accumulating in the settling tank was wasted periodically. Some make-up water was added occasionally to compensate for losses due to evaporation. After an initial period, a detention time of 8.5 days was maintained.

Table 6.11. Operational parameters for Operational Mode III.

Number of birds	226
Number of days	99
Immersion depth of rotor	5.2 cm (2 in.)
Liquid volume	6056 liters (1600 gallons)
Volume of settling tank	1685 liters (445 gallons)
Detention time	8.5 days
Total solids in ODML	$\simeq$1.3%

Table 6.12. Nitrogen losses during Operational Mode III.

Total Nitrogen Losses	kg	%
In settling tank	4.1	9.8
In oxidation ditch	29.8	70.8
In total sytem	33.9	80.6
Total nitrogen input to the system during period of operation = 42.1 kg		

The operational parameters for Mode III are listed in Table 6.11 and the nitrogen balance is presented in Table 6.12. The combined nitrogen loss from the ditch and the tank was about 80%.

Operational Mode IV

In this mode of operation the rotor was not operated continuously. The mixed liquor was permitted to denitrify in the ditch with the rotor off.

The rotor was connected to a time switch which limited the aeration period to a predetermined time interval. Two experiments were made, lasting about three weeks each, in which the rotor was operated for 16 and 12 hours per day, respectively. Wastes from the birds were added continuously, while no overflow was permitted from the oxidation ditch.

Nitrogen balances were computed and are presented in Table 6.13. It was possible to remove up to about 90% of the total nitrogen input to the oxidation ditch by manipulating the aeration period. Higher nitrogen losses occurred when the rotor was operated for 12 rather than 16 hr/day. With the latter aeration period a significant concentration of nitrate-nitrogen (about 40 mg/l) still remained. The 90% loss of total nitrogen observed when the ditch was aerated 12 hr/day suggested that, while nitrification was taking place, the magnitude of nitrate formation was less than that of its removal. Such was not the case when the ditch was aerated 16 hr/day, since the magnitude of nitrification was higher than that of denitrifica-

Table 6.13. Nitrogen balance in an oxidation ditch, Operational Mode IV.[1]

Days	Cumulative Total Nitrogen Input (kg) Rotor Operation		Nitrogen Loss (%) Rotor Operation	
	16 hr/day	12 hr/day	16 hr/day	12 hr/day
2	1.1		34.5	
4		2.1		44.4
6	3.3		49.3	
7		3.7		82.6
9	4.9		51.1	
11		5.9		91.3
13	7.2		53.6	
14		7.5		95.2
16	8.8		58.2	
18		9.6		92.5
20	11.0		62.5	
21		11.2		88.3
23	12.7		69.6	

[1] Amounts of oxidized nitrogen remaining in the mixed liquor at the end of this study were 0 and 2.7 kg, in the 12-hr and 16-hr rotor operations, respectively.

tion and a high concentration of nitrate-nitrogen resulted in the mixed liquor. If this nitrate-nitrogen were subsequently denitrified, the total nitrogen losses would amount to about 90% and would be comparable to those observed when the ditch was aerated 12 hr/day.

From a comparison of these nitrogen balances, it appeared that there was an optimum period of aeration between 12 and 16 hr/day at which the magnitude of nitrification equalled the magnitude of denitrification. If the mixed liquor were aerated for such a period then it should be possible to achieve no accumulation of oxidized nitrogen, while maintaining odorless conditions and accomplishing high nitrogen removals without the aid of additional units for separate denitrification.

The effect of varying periods of aeration on the nitrogen losses from the oxidation ditch are presented in Figure 6.16. Varying degrees of nitrogen removal, in the range of 30 to 90%, can be achieved by suitably adjusting the periods of aeration and *in situ* denitrification in the oxidation ditch.

Summary of Pilot Plant Studies

It is possible to control losses of nitrogen by manipulating the operation of an aeration system such as the oxidation ditch. Ex-

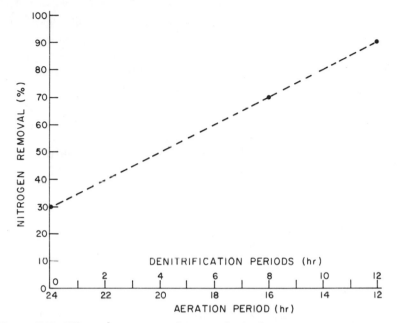

Figure 6.16. Effect of aeration and *in situ* denitrification on nitrogen removal.

Table 6.14. Summary of nitrogen and COD losses in the different modes of operation of an oxidation ditch.

	Loss %	
Mode of Operation	Total Nitrogen	COD
Continuously filling	$\simeq 30$	$\simeq 60$
Continuous flow operation with *in situ* denitrification	$\geqq 30$ depending on the number of denitrification phases	50
Continuous flow operation and recycling of supernatant via settling-denitrification tank	$\simeq 75$	50
Semi-continuous or continuous operation with partial rotor aeration (12 to 16 hr/day)	$\simeq 90$	59[1] 50[2]

[1] At 12 hours aeration.
[2] At 16 hours aeration.

pected nitrogen and COD removals under the different operating conditions are summarized in Table 6.14. Losses were lowest when there was no attempt to cause denitrification in the ditch. The losses could be increased by the following approaches: (1) denitrify the mixed liquor in a separate settling unit without stopping the aeration; (2) denitrify *in situ* by stopping the aeration for an optimal time which is related to the operating condition of the oxidation ditch; and (3) design the aeration system to provide adequate dissolved oxygen concentration initially to achieve nitrification, and have no dissolved oxygen in the rest of the ditch to achieve denitrification. These studies illustrate that oxidation ditches can be managed in several ways to stabilize poultry wastes and control nitrogen. Depending on the mode of operation, it is possible to either conserve as much as 70% or remove as much as 90% of the input manurial nitrogen.

FULL-SCALE OXIDATION DITCH SYSTEMS

System at Houghton Poultry Farm

Description of Operation

In early 1972, a full-scale oxidation ditch was installed at a poultry farm located near Ithaca, New York. The system was installed by the owner, Mr. C. L. Houghton, after observing the pilot plant oxidation ditch. Prior to the installation, waste handling on the farm consisted of liquid collection and anaerobic storage in pits located under the cages. The ends of these pits were connected to form two oxidation ditches. Following the modification, project personnel helped monitor the ditches and identify their performance. The monitoring started in early February, 1972.

The main objective of the ditches was odor control, but their performance as a nitrogen control device was assessed. These oxidation ditches operated as aerated holding tanks, and with the cages above the ditches, input of manure to the stabilization system was continuous. There was a continuous amount of water added to the ditches from a leaky watering system. This leakage necessitated regular removal of the mixed liquor from the ditches. The occasional foaming was controlled by periodic addition of motor oil to the mixed liquor.

Differences in the quantity and frequency of removal of the mixed liquor from the ditches varied the solid retention times, which ranged from 12 to 36 days, and permitted the evaluation of such differences and other operational conditions on the performance of the system.

Methods

Estimates of daily loading of total nitrogen, solids, and COD to the ditches were made by analyzing 24-hr composite samples of bird excreta. Samples of mixed liquor routinely were collected and analyzed for total solids, COD, and the different forms of nitrogen. Mass balances were made to determine the efficiency of nitrogen and COD removals, and the extent of organic nitrogen conversion.

Results

Throughout the study no odors, other than an occasional faint ammonia odor, were evident. The nitrogen concentrations in the ditches are illustrated by the typical curves shown in Figure 6.18. The total nitrogen and COD loadings to the ditches and the removal efficiencies during the different months are noted in Table 6.15. The removal of nitrogen and COD varied from 30 to 33% and 26 to 50%, respectively. The variable removals were a result of varying operational conditions. The nitrogen variations were able to be related to the fundamental factors identified earlier.

The pH of the mixed liquor was always alkaline. Except for a short period about day 260 to 280, when the pH was in the range of 7.6 to 7.9, and nitrites were present in the mixed liquor, the pH value of the liquor was in the range of 8.1 to 8.4. Oxygen input to the system was insufficient to maintain a residual dissolved oxygen throughout the length of the channel. The nitrogen losses were due to ammonia volatilization as well as denitrification.

The major changes in Figure 6.18 at day 80 and about day 270 are the result of different phenomena. At day 80, a major portion of the ditch contents were pumped out to make minor adjustments in rotor height and liquid depth. The conditions and results following day 80 represent those of normal field facilities.

Beginning about day 210 (in August) the waterers for the birds began to leak excessively resulting in a short liquid detention time in the ditches. As a result, a smaller oxygen demand was exerted in the ditches and the oxygen input was adequate to meet the demand. Partial nitrification (nitrite formation) and a decrease in TKN and ammonia nitrogen took place (Figure 6.18). The leakage was corrected about day 275 and the parameters returned to their previous condition.

Figure 6.17. Views of an oxidation ditch for treating poultry excreta on a large commercial poultry farm. Ditch is directly below birds. Motor-driven rotor provides aeration and mixes the liquid. (Photos by R. Loehr)

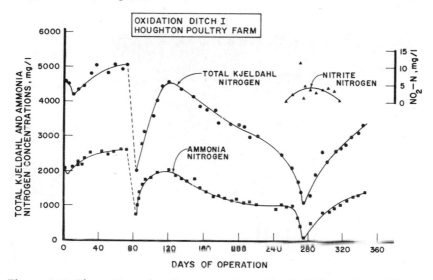

Figure 6.18. Fluctuations in nitrogen content of mixed liquor in oxidation ditch #1 at Houghton poultry farm.

Costs

The investment required for installing the oxidation ditch at the Houghton farm was approximately $10,000, of which $7,200 was for rotors and motors and $2,800 for connecting the ends of the collection pits to form oxidation ditches. Because the collection pits did exist at the Houghton farm, only the cost of converting these pits to oxidation ditches was charged to the oxidation ditch system. Annual costs for depreciation, interest, and repairs were estimated to be 20% of the initial investment. For the 300,000 eggs produced annually at the Houghton farm, these costs were 2/3¢ per dozen. Electrical power to operate the rotors, at 2.17¢ per kWh, amounted to 1.01¢ per dozen. Total cost is thus estimated to be approximately 1-2/3¢ per dozen eggs. While this is probably less than 3% of the cost of producing eggs, it may represent a substantial part of the net income to the operator of the farm. However, in the Houghton case the oxidation ditch substantially reduced complaints from neighbors about odor and may have permitted Mr. Houghton to continue egg production in a location where he might otherwise have been forced to discontinue production or construct, at a substantial cost, a different type of poultry house.

Not included in the above cost estimates are any savings that occurred, such as the elimination of the need to plow down the

Table 6.15. Nitrogen and COD removal efficiencies of the oxidation ditch system at the Houghton Farm.

Month	Average Number of Birds	Detention Time (days)	Total Loading[1] (kg)		Removal Efficiency	
			Nitrogen	COD	Nitrogen	COD
February	14259	29	1294	11950	49	59
March	14114	15	1370	12775	40	56
April	12062	32	1012	9347	37	40
May	12395	36	1203	11105	50	51
June	14777	30	1388	12812	53	42
July	14655	23	1422	13129	46	45
August	14433	16	1410	13020	33	31
September	14450	14	1357	12528	30	37
October	13850	13	1344	12082	58	26
November	14855	12	1395	12879	29	27
December	14740	15	1430	13206	32	36

[1] The average manure loading to the ditches were (g/bird-day) : Total Nitrogen, 3.1; COD, 28.9; and BOD (5-day) 7.9.

manure immediately after spreading and of the need to transport untreated manure considerable distances to minimize odors reaching any neighbors. The elimination of both needs results in a better utilization of manpower which partially offsets the cost of operating the oxidation ditches. A positive but immeasurable benefit of the oxidation ditches is the decrease in concern with odors by both Mr. Houghton and his neighbors and in the time Mr. Houghton spent in defending his operation against nuisance complaints.

System at Manorcrest Farms

Description of Operation

Two oxidation ditches were installed beneath caged laying hens, each receiving wastes from 4000 birds. The mode of operation was continuous flow with supernatant recycle from settling tanks. The system has functioned effectively since installation in August 1973. The two aeration devices installed for the ditches differed in design and oxygenation capacities. The rotor for ditch I had a conventional cage-type design, while that for ditch II was a brush aerator.

Nitrogen Control Efficiency

Figure 6.19 shows the fluctuations of the different forms of nitrogen. The ditches effectively controlled the production of obnoxious odors. Even though 70% of the organic nitrogen in the manure was converted to ammonium, only small amounts of ammonia nitrogen generally were observed suggesting that nitrification was taking place. However, the low levels of nitrites and nitrates indicated that denitrification also was occurring in the systems, and that the input of oxygen was insufficient to prevent nitrogen losses due to denitrification. Ammonia volatilization also was a possibility, although generally very little was available to be volatilized. The Manorcrest data (Figure 6.19) is that resulting from the initial portion of a two-year continuing study.

System at the Mink Farm

A fur animal farm funded by the United States Department of Agriculture needed a waste stabilization system, and personnel associated with this project designed an oxidation ditch based on experiences with poultry wastes. This afforded an opportunity to

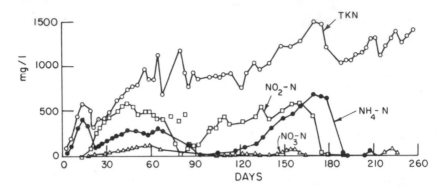

Figure 6.19. Nitrogen content of mixed liquor in oxidation ditch #2 at Manorcrest farm.

verify that the technology developed for one type of waste was
applicable to others.

Results indicated that such a system could be incorporated beneath the confined animals and offensive odors from the manure
eliminated. Aeration of the system was adequate for both odor control and nitrogen conservation.

Nitrates accumulating in the system were removed by stopping
aeration and allowing the liquor to denitrify. Mass balances computed to assess the treatment efficiencies indicated that the system
was capable of removals of about 93, 97 and 46%, respectively, of
nitrogen, BOD and total solids.

From this latter investigation it was concluded that the use of
the oxidation ditch for nitrogen control can be applied to other animal wastes.

Summary of Full-Scale Systems

The discussion of the results from the full-scale studies has been
brief because the different units are under continuing evaluation.
However, the results do indicate that field oxidation ditches can
accomplish varying degrees of nitrogen control while accomplishing waste stabilization and odor control.

Each of the full-scale ditches, *i.e.*, at Houghton, Manorcrest, and
the fur animal farm, was designed for odor control as the waste
management objective. Control of nitrogen to specific levels was
not required and it was not possible to directly apply the modes
of operation and verify the results of the pilot plant studies.

The nitrogen losses from the three field systems and from other

Table 6.16. Nitrogen losses in aerated poultry stabilization units.

Unit	Nitrogen Loss, %
Houghton	29 to 58
Manorcrest	47 to 63
Mink farm	93
Stewart and McIlwain (1971)	66
Hashimoto (1971)	50 to 60
Dunn and Robinson (1972)	70 to 80

systems reported in the literature are summarized in Table 6.16. Although little data on the nitrogen losses from aeration systems treating other animal wastes are available, the losses noted in Table 6.16 would be expected to occur in other aeration systems since the fundamentals of the systems are similar.

SUMMARY OF ENGINEERING STUDIES

The results from the pilot plant and field oxidation ditches clearly identified the level of nitrogen loss that takes place under uncontrolled and controlled denitrification. The studies also indicated approaches that can be used to enhance nitrogen loss where such enhancement is a waste management objective. Equally important, the information also indicated approaches to minimize the nitrogen loss.

It is extremely difficult to conserve all of the nitrogen in animal wastes. Approximately 30% will be lost even in a well mixed and aerated stabilization unit. This amount is less than that identified with the existing methods of waste handling (Chapter 4).

The minimum amount of nitrogen loss will result from the use of completely mixed aeration units that consistently maintain a high dissolved oxygen concentration, i.e., above 2 mg/l. The nitrogen in this mixed liquor will be in the forms of organic, ammonia, nitrite and nitrate. The latter three forms are soluble in water and care should be taken to minimize losses of these forms when the oxidation ditch mixed liquor is applied to the land. The greatest benefit will result when the soluble nitrogen can be utilized by the crops shortly after the mixed liquor is applied.

If the mixed liquor is stored before application to the land, the storage should be under well aerated conditions. Otherwise the oxidized nitrogen, nitrite and nitrate will be denitrified, lost from the mixed liquor, and will negate the nitrogen conservation achieved in the initial oxidation ditch.

To achieve the minimum nitrogen loss, the oxidation ditch should be operated as a well-aerated holding tank with the contents only removed when they can be used effectively in crop production.

The maximum nitrogen loss will occur under conditions that favor ammonia volatilization and denitrification. These conditions occur when the oxidation ditch is poorly mixed, when aeration is intermittent, when nitrified wastes are held under nonaerated conditions, and when the dissolved oxygen concentration in the mixture remains low, i.e., below 1 mg/l.

Achieving the desired nitrogen control can be done by using existing knowledge of design and operation. As noted in the introduction to this chapter, the decisions on what nitrogen control approach is appropriate depends upon economic, legal, manpower and other constraints imposed upon the producer. The basic decisions are to achieve a) maximum nitrogen conservation and use in crop production as the wastes are disposed of on the land, or b) nitrogen loss to the level that the nitrogen remaining in the wastes is consistent with the land application guidelines. In the latter case, excess nitrogen loss to the environment can be minimized.

Nitrogen conservation, rather than loss, will be the goal of most livestock producers. The increased cost and possible uncertain availability of inorganic fertilizers are causing considerable interest in the nutrients in animal manures. Large-scale poultry production is such that some producers are not engaged in crop production and therefore may not have adequate land for prudent disposal. In such cases, a degree of nitrogen removal may be not only desirable but necessary.

The studies described in this section provide information on the technical alternatives that can be used to achieve the degree of nitrogen control that is desired, whether the emphasis be conservation or loss. The results of the studies also provide information on the forms and quantities of the nitrogen that will be in the wastes disposed of on the land whatever alternative is chosen. This latter information, when integrated with the material found elsewhere in this report, is useful in identifying the potential environmental hazards that may occur when the stabilized wastes are disposed of on the land.

The pilot plant and field oxidation ditch studies complement the previous laboratory studies. These larger systems do not operate at the optimum pH or temperature values and the rates of nitrification and denitrification obviously are reduced. However, the detention times of the systems are long and compensate for the reduced microbial reaction rates.

CONCLUSIONS

Proper stabilization and management of livestock wastes in liquid treatment systems can eliminate unpleasant odors while providing management of nitrogen- and oxygen-demanding components of the wastes. When wastes are disposed on land, it is unnecessary to remove BOD and COD to the degree that would be necessary if the wastes were discharged to streams.

Given adequate cropland, disposal of livestock wastes can be integrated with crop production, in which case it may be desirable to conserve nitrogen in the waste prior to its application to the land. Where land is limited, nutrients such as nitrogen may have to be removed so the assimilative capacity of available soil is not exceeded.

To achieve control over the levels of nitrogen in wastes it is necessary to understand the fundamentals of possible approaches and the performance of feasible treatment systems. Nitrification-denitrification approaches are feasible and can be incorporated in liquid, aerobic animal waste stabilization systems. Since nitrification is caused by actively nitrifying organisms, estimates of the amount of these organisms and of the quantity of inorganic nitrogen in the system should be used to express design factors for nitrification rather than base the design of nitrification systems upon estimates of heterotrophic organism mass of activity and BOD. This study described an attempt to (1) understand the factors influencing nitrification reactions and develop a procedure for estimating the active mass of nitrifying organisms; and (2) examine the feasibility of using oxidation ditches for achieving nitrogen control.

It is difficult to operate a waste stabilization system without losing some nitrogen. Losses due to ammonia volatilization can be minimized by nitrifying waste. Anoxic pockets and facultative conditions prevailing at the floccular level of a nitrifying mixed liquor can lead to denitrification of the oxidized nitrogen and preclude the possibility of conserving all the nitrogen entering the system. If conservation of nitrogen is desired, such losses should be minimized by maintaining adequate levels of oxygen throughout the system. These results show that it is possible to achieve different degrees of nitrogen removal (ranging from 30 to 90%) with different modes of oxidation ditch operation by controlling the time of effective nitrification and denitrification.

COD removals of 50 to 60% were achieved in all operational modes of the ditch. Although such removals are not required for disposal of treated wastes on land, the corresponding reduction in odor and in the quantity of polluting substances has environmental benefits.

ACKNOWLEDGMENTS

The following individuals were involved in the details of these engineering studies. The specific contribution of certain individuals is noted.

Y. D. Joo, Research Technician

J. H. Martin, Jr., Research Specialist (field studies)

W. D. Morse, Graduate Student

T. B. S. Prakasam, Senior Research Associate

E. G. Srinath, Research Associate (active mass studies)

G. Wong-Chong, Research Specialist (laboratory nitrification studies)

Principal Author

Raymond C. Loehr

REFERENCES

Anon. (1971). "Standard Methods for the Examination of Water and Waste-water," 13th ed., American Health Assn., New York.

Anon. (February 14, 1974). Effluent Guidelines and Standards, Part 412, Feedlots Point Source Category, Federal Register, Volume 39, pp. 5704–5710.

Anthonisen, A. C. (1974). "The Effects of Free Ammonia and Free Nitrous Acid on the Nitrification Process," Ph.D. Thesis, Cornell University, Ithaca, New York.

Dunn, G. G. and Robinson, J. B. (1972). "Nitrogen Losses through Denitrification and Other Changes in Continuously Aerated Poultry Manure," Proc. Agric. Waste Management Conf., pp. 545–554, Cornell University, Ithaca, New York.

Hashimoto, A. G. (1971). "An Analysis of a Diffused Air Aeration System Under Caged Laying Hens," Paper NAR-41–428. Amer. Soc. Agr. Engr., St. Joseph, Michigan.

Jeris, J. S. (1967). "A Rapid COD Test," Water and Waste Engineering 4, pp. 89–91.

Loehr, R. C., Prakasam, T. B. S., Srinath, E. C., Scott, T. W. and Bateman, T. W. (1975). "Design Parameters for Animal Waste Treatment Systems—Nitrogen Control," EPA Project #S800767, ROAP 21AYU, TASK 02, Environmental Protection Agency, Washington, D.C.

Loehr, R. C. (1974). Agricultural Waste Management. Chapter 2, Academic Press, New York.

Loehr, R. C. and Johanson, K. J. (1974). "Phosphate Removal from Duck Farm Wastes," Journal Water Pollution Control Federation, Vol. 46, No. 7, pp. 1692–1714, Washington, D.C.

Prakasam, T. B. S., Srinath, E. G., Yang, P. Y., and Loehr, R. C. (1972). "Evaluation of Methods for the Analysis of Physical, Chemica, and Biochemical

Properties of Poultry Waste," presented at the Winter ASAE Meeting, Chicago, Illinois.

Prakasam, T. B. S., Loehr, R. C., Yang, P. Y., Scott, T. W., and Bateman, T. W. (1974). "Design Parameters for Animal Waste Treatment Systems," EPA-660–2–74–063, Project Number S800767, ROAP/TASK 21 AYU-01, Environmental Protection Agency, Washington, D.C.

Srinath, E. G. and Loehr, R. C. (1974). "Ammonia Desorption by Diffused Aeration," *Journal Water Pollution Control Federation*, Vol. 46, No. 8, pp. 1939–1957, Washington, D.C.

Stewart, T. A. and McIlwain, R. (1971). "Aerobic Storage of Poultry Manure," *Livestock Waste Management Pollution and Abatement*, Publ. PROC-271, pp. 261–263, Amer. Soc. Agr. Engr., St. Joseph, Michigan.

APPENDIX A

Estimation of Active Nitrifying Mass

Experimental observations described in the body of this chapter indicate that it is possible to estimate the active nitrifying mass using controlled conditions. The actual procedure for estimating the active mass is as follows:

1. Take two aliquots of the mixed liquor sample; centrifuge both aliquots to separate the solids. Discard the supernates. Replicated samples are advisable unless one is confident that a single sample is representative.

2. Wash both sediments with physiological saline (0.9% NaCl solution) by repeated resuspension and centrifugation. Discard the supernates. The samples should be washed until there is no measurable ammonia, nitrite, or nitrate in the supernatant. Generally three washings will suffice.

3. Transfer all of the washed solids from the centrifuge tubes, one into each of the following:

    (a) 100 ml of substrate (solution containing 100 mg NH_4^+-N per liter in phosphate buffer, 0.05M, pH = 7.8) in a 250-ml Erlenmeyer flask. This solution will provide information on the reaction rate of *Nitrosomonas*. To prevent further oxidation of NO_2^- to NO_3^-, potassium chlorate ($KClO_3$) at a concentration of 10 mg/l could be included in the substrate. This step may be omitted in cases where high nitrifying activity is not encountered.

    (b) 100 ml of substrate (solution containing 100 mg NO_2^--N per liter in phosphate buffer, 0.05M, pH = 7.8) in a 250-ml Erlenmeyer flask. This solution will provide information on the reaction rate of *Nitrobacter*.

4. Place a magnetic stir bar in each of the flasks and keep the reaction mixture well stirred by placing the flasks over magnetic stirrers.
5. Determine NO_2^-–N content of the reaction mixtures at intervals of 1 hour up to 6 hours, and after 24 hours.
6. Plot the graph of concentration of NO_2^-–N vs. time.
7. Determine the value of the slope (oxidation rate-k) of the NO_2^-–N vs. time curve. Average all of the k values of any replicates to obtain an average rate for the respective samples.
8. Determine the TKN content of the suspended solids in the reac-

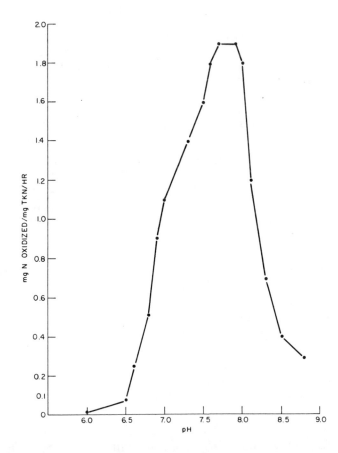

Figure 6.20. Effect of pH on the specific nitratification activity of *Nitrobacter*.

tion mixtures. Average the TKN values of any replicate samples to obtain an average value for the respective samples.

9. Determine the specific activities of NH_4 and NO_2 oxidation by dividing the oxidation rate (#7) by the TKN of the suspended solids in the system (#8). Let these be $A_{NH_4}{}^+$ and $A_{NO_2}{}^-$.

10. Active nitrifying mass in the mixed liquor sample is calculated as follows:

(a) *Nitritifying activity:*

$$\text{Active mass of nitritifiers} \atop \text{(mg TKN/l)} = \frac{A_{NH_4{}^+}}{\text{Specific activity of }\textit{Nitrosomonas} \atop \text{at pH 7.8 (such as 0.6 from Figure 6.12)}} \cdot \text{TKN of SS}$$

(b) *Nitratifying activity:*

$$\text{Active mass of nitratifiers} \atop \text{(mg TKN/l)} = \frac{A_{NO_2{}^-}}{\text{Specific activity of }\textit{Nitrobacter} \atop \text{at pH 7.8 (such as 1.9 from Figure 6.20)}} \cdot \text{TKN of SS}$$

Using the pH-specific activity relationships (Figures 6.12 and 6.20), the oxidation rates obtainable in a system maintained at different pH values could be predicted if the oxidation rate at any one pH value is known.

Implicit in this approach is the assumption that the conditions of the test do not result in any stress such as inhibition or oxygen limitation.

7

Sociological Investigations of Three Watersheds

Chapter 7

SOCIOLOGICAL INVESTIGATIONS
OF THREE WATERSHEDS

Now, more than ever before, nature appears to have acquired expressive meaning for the American people rather than being, as before, merely an object for consumptive use and conquest.

Hendee, Gale, Harry: 1969

INTRODUCTION

Society and Its Management of Water Resources

Although the physical, chemical, and biological aspects of man's interactions with the environment have received considerable attention, a point of view seldom articulated is that the decisions and activities of individuals and their actions as groups are of primary importance in modern agricultural and resource management. It is increasingly recognized, however, that this sociological perspective is essential in solving environmental problems.

This perspective is expressed by all social groups to varying degrees depending upon their position and perception of environmental problems. Many of the views expressed pertain to different segments of American society, but all are considered appropriate for the geographic focus of this study, namely central New York State.

Resolution of the conflicting use of resources within watersheds requires the support of individuals along with appropriate organizational and institutional response. The population of a watershed constitutes a variety of social groups; and although they share common water resources, this fact seldom provides the basis for a social structure organized to define standards for effectively controlling water quality. The ability to cooperate, develop new legislation, levy taxes, or enforce new standards depends on other aspects of the social system, such as jurisdiction of local government, political boundaries, and historical precedence.

Local political boundaries are notable in fragmenting jurisdiction and authority when issues cross or transcend these boundaries. This is especially the case with management of water resources, where the lack of politically organized units encompassing particular watersheds and inadequate social and economic resources make it difficult to implement effective policies at a watershed level. Efforts to define, protect, and maintain the quality of a water resource necessarily involve many components of the social system, such as individuals, special interest groups, peer groups, local political units, state and national agencies, and other institutions formed to deal with problems of water quality.

Many of the social decisions arising from the interactions of groups with competing interests are translated into laws or regulations which have the object of ensuring that everyone is affected fairly. For such regulations to be effective, they must be based on known cause-and-effect relations, be enforceable, and receive public support.

If there is public pressure for reduction of nutrient inputs to waters before cause-and-effect relations have been adequately quantified, remedial action may create new difficulties. For example, the inhabitants of a rural community may be worried about the degradation of a small local stream. The cause could be attributed to leakage from residential septic tanks. The remedy may consequently be to install a sewer system and a sewage treatment plant. However, it is possible that the discharge from the proposed plant could be more deleterious for the stream than the combined effects of the dispersed septic tanks because of the relatively large volume of wastewater discharged at one point. Even if the standard of treatment provided by the sewage treatment plant is high, the amount of dilution in the stream is not likely to be adequate to accommodate the plant discharge of nutrients without a marked change in the quality of the stream.

Once the nature and scope of environmental problems have been defined, appropriate management alternatives must provide for the interface of functions and the coordination of activities of institutions operating with different authority, jurisdiction and resources. A level of compliance with regulations may be achieved through education or the use of positive or negative sanctions, but public support will be determined by individual values, attitudes, and overall perceptions of the situation.

Efforts to modify patterns of behavior to protect and maintain the quality of water resources will be effective only if they take into account the factors that motivate and influence individuals and groups. To significantly change long-established patterns of behavior that affect the environment, the ways in which new patterns can be supported and reinforced at individual, local, institutional or national levels need to be explored.

Sociological Research on Environmental Problems

Sociological research can be utilized to determine the public's attitudes and concerns regarding water quality, to determine the level of support people will lend to improve the aquatic environment, and to examine the available means by which they can either express their concerns or undertake remedial action. Often motivation exists, but the social organization for action is deficient.

In this context, sociological research can indicate and help to define the nature of public problems, as well as identify alternative solutions and modes of implementation. It can point out how different patterns of social organization can either constrain or facilitate the solving of problems and the conditions essential for effective action. Sociological studies can provide insight into the processes of decision making, particularly indicating the relationships between issues people perceive and the capacities of institutions or organizations to implement proposed solutions.

Environmental sociology as a field of study is still very new, being broadly defined as the study of the behavior of social groups as they affect and are affected by environmental problems. Public opinion polls concerned with the environment were first conducted in 1965 (Erskine, 1972a, 1972b), with research beginning shortly thereafter. At present a body of knowledge is emerging, although its empirical foundation is still limited relative to other areas of sociology, and a separate conceptual basis is as yet almost nonexistent.

Environmental sociology research has considered several categories utilizing various levels of analysis, such as individuals, groups, communities, regions, and state or national levels.

Studies of individuals have tried to determine values, attitudes, and perceptions regarding environmental problems (Dasgupta, 1967; Peterson, 1971; Simon, 1971; Murch, 1971; Heberlein, 1974; Erskine, 1972a, 1972b). The interview questionnaire has been the primary tool employed in these studies, along with the use of opinion polls. In a few cases, efforts have been made to construct a behavioral model based on findings from questionnaires and polls. The study by Gore *et al.* (1972) produced a provisional model of this nature.

At the group level, the roles and activities of various special interest groups in solving environmental problems have been studied. Examples are citizens' associations, environmental protection groups, and property owners (Finley and Hickey, 1971).

Studies at the community level have investigated the community's capacity to solve its environmental problems through institutions, information, human skills, and financial resources (Ireson, 1972). Various social organizations, such as water development districts, have been analyzed primarily through case studies (Wilkinson, 1966; Finley and Baker, 1972; Citizens Advisory Committee on Environmental Quality, 1972).

At the regional, state, and national levels, efforts have been made to understand the issues of planning, policy formation, and the roles of various governmental bodies (Capener *et al.*, 1974; Holcomb *et al.*, 1974; Haskell and Price, 1973). Political scientists have also investigated this area, concentrating on the power relationships in the environmental policy-making process (Smith *et al.*, 1974; Borton and Warner, 1970; Davies, 1970; Ridgway, 1971; Thompson, 1972).

At least one effort has synthesized some of the above considerations into a model which relates water resource policy variables to social goals (Utah Water Research Laboratory, 1971). This is notable, given that the world model presented in *Limits to Growth* (Meadows *et al.*, 1972) ignored both human and social organization factors.

OBJECTIVES OF THE
SOCIOLOGICAL INVESTIGATIONS

Given the lack of knowledge of sociological aspects of water quality, an overall purpose was to establish groundwork that might be useful in future research. The more specific objective of this study was to determine the public's attitudes and concerns regarding stream and lake water quality. To meet this objective, it was necessary to determine the characteristics of the population, their uses of water, awareness of pollution, willingness to achieve solutions, attitudes toward regulation of water quality, and preferred means of enforcing regulations.

WATERSHEDS SELECTED FOR STUDY

The research initially concentrated on the population in the Fall Creek watershed (Figures 7.1 and 7.2), the physical aspects of

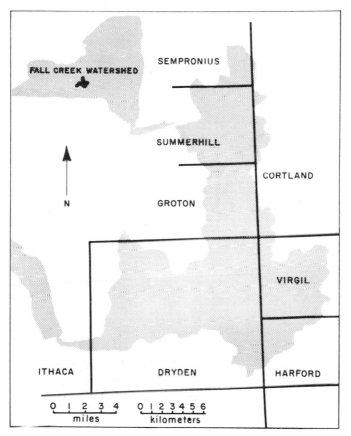

Figure 7.1. Outline map of Fall Creek watershed showing boundaries of townships.

which are described in Chapter 3. To provide a useful comparison, the populations of two other watersheds were studied. These are the Owasco Lake watershed (Figure 7.3) and the Canadarago Lake watershed (Figure 7.4). The three watersheds are similar, being predominantly rural and agricultural, and located in central New York State.

Owasco Lake is approximately 25 miles southwest of Syracuse, New York. Both the city and the township of Auburn are located at its outlet. Summer residents and farm families live on its shores. Research supported by Cayuga County and its Environmental Management Council has been conducted on the factors affecting lake water quality for the purpose of making recommendations for water quality maintenance and improvement (Oglesby *et al.*, 1973).

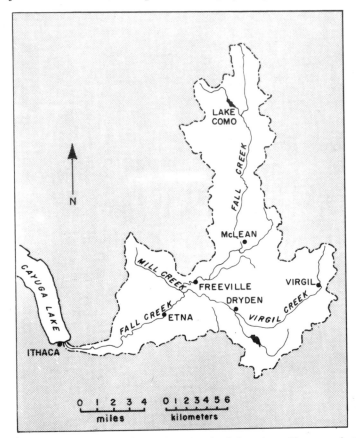

Figure 7.2. Outline map of Fall Creek watershed showing villages and streams.

The third watershed, Canadarago (located equidistant between Syracuse and Albany just south of the Mohawk River), was chosen for several reasons. The New York State Department of Environmental Conservation (Hetling, 1971) was involved in an extensive study of the sources of nutrients to the lake along with a joint plan with the Village of Richfield Springs to build a tertiary sewage treatment plant. The addition of the sociological study to the substantial biological data seemed advantageous. Also, the interest and activi-

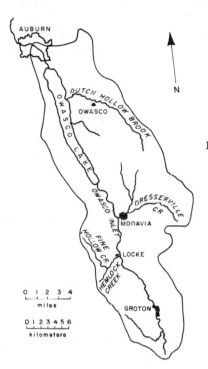

Figure 7.3. Outline map of Owasco Lake watershed. See Figure 2.2 for location of Owasco relative to the other Finger Lakes.

Figure 7.4. Outline map of Canadarago Lake watershed.

ties of the Canadarago Lake Association in management and preservation of the lake was a positive factor.

INVESTIGATIONAL APPROACH

Questionnaire*: Construction and Sampling Procedure

The questionnaire as a date-gathering instrument has traditionally been the primary technique employed by behavioral scientists to measure attitudes of large populations. In this study, a questionnaire provided field interviewers the means of collecting data on the social characteristics of the watershed populations, their uses of water, and their attitudes about responsibilities for pollution problems. The form and content of the questionnaire were developed with the cooperation of representatives of various local decision-making groups, such as county planning boards, health departments, and the Cooperative Extension Service. The questions were similar to those used by various national polling organizations (e.g., Erskine, 1972a, 1972b) and by other studies which have addressed problems of water quality. The questionnaire was pretested in a comparable adjacent watershed, which also provided training for the graduate student interviewers.

Details of the watersheds were identified using U.S. Geological Survey maps which were subdivided into grid squares of one kilometer. Squares with high residential concentration were considered separately in sampling by using more detailed maps. Remaining squares were numbered and one-sixth of these were selected using random numbers. All heads of household (either male or female) in the selected squares were interviewed. All houses were numbered, and one-sixth of the households in each watershed were sampled as above. Table 7.1 notes the distribution of the interviews in the three watersheds.

Background Information on Populations

General socioeconomic information on the residents of the three watersheds is summarized in Table 7.2. Although the characteristics of the residents differed somewhat among the three watersheds, no significant differences in attitudes or behavior among the three were

---

* Copies of the questionnaire are available from the Department of Rural Sociology, Cornell University, Ithaca, New York 14853.

Table 7.1. Summary of interviews in the three watersheds.

	Fall Creek	Owasco Lake	Canadarago Lake
Area of watershed (square kilometers)	336	548	197
Proposed number of interviews:			
Rural area	294	302	68
Higher density area	273	376	216
Total	567	678	284
Number of interviews not completed	144	142	84
Number of interviews completed:			
Rural area	237	248	46
Higher density area	186	288	154
Total	423	536	200
Total interviews for all three watersheds			1159

found. The respondents displayed a remarkable consistency in their responses. Therefore, the results of all three surveys are combined for the following discussion.

Residents in the watersheds tended to make considerable use of the water. About 40% of the residents lived within one kilometer of the lake or stream. Two-thirds of the respondents engaged in water-based recreation more often than once a month during the summer; one-third of this group did so more often than once a week. Swimming was the most popular activity (36%), followed in preference by fishing (30%) and boating (24%). One-third of those

Table 7.2. Basic information on the residents of the three watersheds.

Characteristics of Residents	Percentage in Survey According to Watershed		
	Fall Creek	Owasco Lake	Canadarago Lake
Have resided more than 5 years in same town	58	76	72
Home ownership	86	86	88
Have completed high school	77	71	66
Rural resident (i.e., not living in built-up area)	52	63	23
Retired	14	23	22
Median income	$9,750	$10,215	$9,180

interviewed owned at least one boat. Approximately one-third of the sample had residences with frontage on a stream or lake. More than half the sample (58%) believed that property values were affected by water quality.

A high degree of subjective awareness of water quality was indicated by the survey; seven out of ten persons reported having seen signs of pollution in the stream or lake, such as detergent foam or trash. Half had not noticed any changes in the quality of the water during their time in residence, but a third said water quality had deteriorated. Three out of four persons interviewed indicated they believed the water was unsafe to drink. One out of five persons felt that it was unsafe to swim in or eat fish from the stream or lake.

PRESENTATION OF FINDINGS

The Watershed as a Unit of Investigation

To determine whether the whole watershed as an area was appropriate for the purposes of sociological research, the sample was divided into two groups: those residents living within one kilometer of the water resource (47%) and all others (53%). When the attitudes and behavior of these two groups were compared, there were no significant differences. Proximity to the water did not influence attitudes or behavior.

A similar comparison was made of residents of incorporated villages (48%) and other inhabitants of the watersheds (52%). Although these two groups differed in such social characteristics as education, time in residence, media participation, and frequency of water recreation, none of these differences were related to attitudes regarding water quality.

In sum, water quality perceptions and attitudes were not affected by location of residence. These findings strengthen the rationale for treating the watershed as a conceptual unit for research purposes; it is the most feasible approach ecologically and there are no serious constraints sociologically.

Questions Important to the Research

Several specific questions were particularly relevant to the investigation:

1. How sensitive are people's perceptions of water quality?
2. How serious do they think water quality problems are?

Figure 7.5. Typical landscape in the Fall Creek watershed. Note the rural community, the dairy farm and the variety of rural land uses in the background. (Photo by G. Casler)

3. Are they willing to act to protect water quality?
4. Are they willing to pay for improvements?
5. How do they believe water quality can be best maintained?

Each of these questions is considered below. In interpreting them it is important to note that there is an "apparent" inconsistency in what respondents perceived as standards of water quality and what physical measures of water quality revealed. In addition there were "apparent" differences in what people perceived about standards of water quality and how they used the water, *i.e.*, for swimming, fishing or boating. Finally there are "apparent" differences in how strongly people feel about the seriousness of the problems of water pollution and their actual willingness to pay for or otherwise support control measures.

Sensitivity to Water Quality

Respondents believed they were sensitive to changes in water quality and were willing to assign different values to the water depending on its use for recreational purposes. In the 1973 Canadarago Lake study (Saint *et al.*, 1974), respondents were asked to rate lake water from "very poor" to "excellent" for swimming, boating, and fishing. By assigning values to their responses (very poor—1; excellent—5), the quality of the lake's water was assessed with regard to its utility for these recreational activities. From the summarized data in Table 7.3 it appears that the lake water was considered more suitable for boating than for swimming or fishing. Since boating depends very little on the quality of water, this seems reasonable. The data, however, suggest that the respondents were less insensitive to changes in water quality. During the summer, there was an increase in the clarity of the lake's water as a spring algal bloom subsided (Table 7.3). While the water quality changed dramatically, there was only slight change in the visual perception indices. With respect to boating activities the perception of water quality was likely influenced more by heavy boat·traffic, a factor which is not a property of the water itself.

One difficulty in detecting changes in water quality is that the appearance of a body of water may change according to the lighting and wind conditions and other external factors independent of actual water quality. This difficulty was further confirmed by showing respondents in the Owasco Lake Study photographs of lakes which purported to show different degrees of water quality. However, the results were inconclusive as there was evidence that the respondents were influenced by scenery and objects in the background, such as buildings, cattails, or beautiful woods, as much as by the appearance of the water.

Seriousness of Issues of Water Quality

Watershed residents generally reflected the concern for water quality problems demonstrated by New York residents in a 1970 Harris poll (Harris, 1971). The poll found that state residents believed that pollution was the most serious statewide problem, with water pollution being the most serious kind.

Half the respondents in the three watersheds considered that water quality was a serious problem. Almost half of the residents (46%) in the Fall Creek and Owasco Lake watersheds had heard about local water conditions potentially hazardous to public health, as opposed to only one out of eight persons in the Canadarago Lake watershed. One-third of all respondents believed that water quality was deteriorating.

Respondents were virtually unanimous in their belief that the quality of the environment should be improved (98%). Eight out of ten respondents in the Owasco and Canadarago Lake watersheds stated that they were optimistic that this could be done, whereas in

Table 7.3. Subjective ratings by lakeside residents of suitability of water quality for recreation according to calendar periods in 1973.[1]

Period of Assessment for Lakeside Residents	Average Rating[2]			Associated Reading of Transparency[3]	
	Swimming	Boating	Fishing	Date	Depth Visibility (meters)
Prior to July 10	3.23	4.08	3.20	June 21	0.6
				July 6	0.6
July 11–August 7	3.52	4.14	3.26	July 19	4.3
				August 2	5.8
August 8 and after	3.61	3.74	3.44	August 16	5.3
				August 30	5.3

[1] Ratings used were 1—very poor; 2—poor; 3—fair; 4—good; 5—excellent. Associated transparency measurements are also given, based on the Secchi disc. (See Figure 2.2 for explanation.)
[2] The results in this study represent a total enumeration of all lakeside residents.
[3] Department of Natural Resources, Cornell University.

Fall Creek less than half the respondents (48%) shared this optimism.

Willingness to Act

In support of this opinion, respondents were asked if they would be willing to take deliberate action, such as requesting a neighbor to stop polluting or reporting him to an official. Three out of four persons said they would take some action themselves if they knew a neighbor was polluting a lake or stream. Half of these stated that action would be taken by involving a public official.

These actions are not carried out to the degree implied, however. One-third of the residents were aware of specific violations regarding sewage, but only 6% had complained to a public official. Given the multitude of agencies with responsibilities for water quality and the overlapping jurisdictions of many of these agencies, the average person does not know whom to contact nor how authorities should be informed. In addition, there is evidently reluctance to strain relations with a neighbor.

The watershed residents' stated willingness to act has increased in recent years. In the 1971 Canadarago Lake study, three out of four persons said they would take action if a neighbor was polluting, but by 1973, nine out of ten persons expressed this predisposition.

Firm support was indicated for public intervention to protect water quality. There was strong backing for government monitoring and subsidy of pollution control. There was also considerable support for government control of land use practices (79%); even three out of four farmers favored it. Most respondents were selective in their preferences for the use of legal sanctions against polluters; three-fourths favored stronger actions against industrial polluters, but preferred more lenience for farmers.

Willingness to Pay for the
Preservation of Water Quality

A willingness to share the costs of preservation of the aquatic environment was expressed by a majority of respondents. Three-fourths of the residents stated that all available pollution control techniques should be applied, even considering the cost. Since one-third believed the costs of pollution control were high, this response was unexpected. More specifically, two-thirds of the respondents stated they would pay an average of 8.5¢ more per half gallon of milk or per dozen eggs if it would permit farmers to produce these products without polluting the environment. This shows some concern with the nature of the farmers' problems of handling wastes, as discussed in Chapter 5.

People also believed certain groups should receive more financial assistance than others in their efforts to combat water pollution. One-third supported aid for businesses, one-half thought households should receive aid, while two-thirds supported help for farmers. (See later discussion on farmers, p. 292.)

In sum, people believe that more money should be spent to control water pollution and were personally willing to share the costs. They also believed that certain groups should have more financial assistance than others, but there was no consensus on the source of these funds. Increases in income taxes, property taxes, and food prices, reordering of governmental budget priorities, and special water use taxes all received about equal stated support. However, it was evident that people supported fair distribution of costs as well as clear delineation of expenditures.

Maintenance of Water Quality

A majority of respondents believed that existing pollution control was ineffective (72%) and that standards were too lenient (59%). They also believed that regulations were unfair to the extent that new homes had much more stringent requirements for sewage disposal than older homes, and dairy farmers were more strictly regulated than homeowners.

Most respondents had specific ideas for increasing the effectiveness of pollution controls. Three-fourths said that there should be some public control of land use practices in the watershed and that those causing pollution should be prevented by law. These results show a clear mandate for more governmental regulation of water quality.

Opinions were expressed, however, that governmental intervention and control should be carried out at the local level, and that water pollution control planning should follow natural watershed boundaries rather than political lines. Often there was no obvious visible structure or organization for people to approach or work through. In response, people favored the creation of special watershed districts as administrative units to address problems of water quality. Over half the respondents believed that water quality control was a local responsibility, while one-fourth favored the state as the appropriate administrative unit.

The watershed surveys generally confirmed the findings of various national surveys (Erskine, 1972a, 1972b). On both national and watershed levels, a high degree of concern with water quality and willingness to pay for water pollution abatement were expressed. Both national and watershed respondents believed that sewage disposal was a more important problem than farm runoff of fertilizers or insecticides.

STUDIES OF COMMUNITIES AND LOCAL GROUPS IN RELATION TO WATER QUALITY

Socioeconomic Topography

As a result of the investigations reported above, it was concluded that watershed residents were concerned about water quality and willing to act to preserve it. However, there was confusion as to how violations should be reported and which persons or agencies were responsible.

Having recognized this situation, the study subsequently investigated the extent to which local communities were prepared to deal with problems of water quality. As this investigation progressed, it became evident that in addition to the physical topography of the watershed, there exists a "socioeconomic topography." The watershed's socioeconomic topography, which encompasses the social organization and social characteristics of the watershed residents, can facilitate or constrain efforts to improve water quality in much the same way as do physical features, such as soil type, vegetative cover and slope. The Fall Creek watershed provides an example of what is meant by "socioeconomic topography" and how it influences social organization and the capacity of local communities to confront environmental problems.

Communities in the Fall Creek Watershed

The Fall Creek watershed includes six villages, eight townships, and parts of three counties (Figure 7.2). There is considerable diversity in the distribution of population and socioeconomic characteristics among these townships. This is apparent from Table 7.4, in which seven of the townships are represented. The eighth, Ithaca, contains the City of Ithaca (population 41,846 in 1970) and is located at the mouth of Fall Creek. Because of its much larger size, it was omitted from the table.

The variation observed in this data suggests that the nature of nu-

Table 7.4. Socioeconomic characteristics of towns in the Fall Creek watershed.

Characteristics	Townships						
	Dryden	Groton	Summerhill	Virgil	Harford	Homer	Sempronius
Area (sq km)	243	128	67	123	62	131	75
Population (1970)	9,770	4,881	670	1,692	748	6,480	649
Population density (per sq km)	40.2	38.1	10.0	13.8	12.1	49.5	8.7
Population change, 1960–70 (%)	32.9	9.2	0.4	19.2	17.8	12.7	18.4
Population living in same house as of 1965 (%)	45.6	61.4	64.7	80.5	54.8	64.2	58.2
Population living in different house from 1965 (%)	21.3	19.5	19.7	4.5	19.3	26.0	10.1
Population employed inside the county (%)	91.6	61.7	39.1	88.8	65.0	85.1	79.4
Population employed outside the county (%)	8.4	38.3	60.9	11.2	35.0	14.9	20.6
Gini coefficient[1]	.341	.314	.335	.285	.293	.391	.310
Percent of employed persons in agriculture	5.0	7.6	25.2	13.8	12.5	7.6	14.6
Percent of employed persons in manufacturing	21.9	48.0	39.5	42.3	36.0	32.0	43.0
Percent of employed persons in finance[2]	20.4	12.2	10.5	14.6	5.1	3.1	14.9
Percent total population in families with less than poverty level income	6.8	11.9	15.8	4.9	4.7	10.1	17.7

[1] The Gini coefficient is a measure of the equality of income distribution within a given population. Since a perfectly equal income distribution would yield a Gini coefficient of 0.00, the greater the coefficient, the more unequal the distribution of income within that population.

[2] Finance includes those employed in wholesale and retail trade, finance, insurance, real estate, and business and repair services.

trient inputs to the stream and its tributaries may vary with changes in socioeconomic topography from town to town. For example, the nature of that input may differ depending on the density and growth rate of the population as well as production activities in each town. For example, Dryden has a relatively high population density and growth rate; hence, the nutrients it contributed to the stream were primarily from sewage. In smaller towns, such as Summerhill, where one-fourth of the employment is agricultural, farms are more likely than sewage to be sources of nutrients in the water.

Responsibility for regulation of this diverse land and water use is divided among several administrative units with frequent overlapping of jurisdiction. Three county health departments legislate and supervise their own standards for installation and use of septic tanks. There are three county planning offices, each with its own perspective. In addition, there are three county boards of supervisors and six village councils. Organizations outside the watershed, such as the Cayuga Lake Basin Board, the New York State Department of Environmental Conservation, and the U.S. Department of Agriculture Soil Conservation Service, all claim some responsibility for the planning and control of land and water use within the watershed.

Financial resources available to a community also limit its capacity to solve local problems. As a hypothetical example, assume that sewage from a town is contributing undesirable amounts of nutrients to a stream, and that the remedy is installation of a sewage treatment system for which funds have to be provided. The funds may be raised several ways: borrowing, local taxation, or state and federal assistance. An illustration of the potential for raising local taxes is shown in Table 7.5, in which it is evident that towns like Summerhill and Sempronius with the lowest populations and assessed value of property pay vastly higher proportional property taxes. Two conclusions follow. First, towns with low assessed property values already have a relatively high level of property taxation with poor prospects for further increases, and conversely. Second, towns with larger resources will more easily meet new financial burdens. For example, an investment of $500,000 in any given project would obviously be easier for Groton than for Virgil. If the investment were paid off over ten years at 7% interest, the average yearly cost would be $69,250. This would necessitate increases in tax revenues of 0.4% for Groton as compared to 1.4% for Virgil.

Apart from financial resources, social organization and human

Table 7.5. Relative financial effort among watershed towns.[1]

Towns	Assessed Value of Real Property (1)	Property Tax Revenue (2)	Percent (2) is of (1) ("Effort")
Dryden	$38,247,500	$251,179	.7
Groton	16,289,216	103,451	.6
Summerhill	705,267	20,876	2.9
Virgil	4,903,550	52,965	1.1
Harford	2,355,666	31,949	1.4
Homer	23,766,600	63,834	.3
Sempronius	1,443,487	48,713	3.4

[1] Source: "Special Report on Municipal Affairs by the State Comptroller for Local Fiscal Year Ended in 1971." Legislative Document No. 96, transmitted to the State Legislature (New York), March 1, 1972.

resources are also important. For example, in 1965 the Village of Dryden (population 1,490) calculated that the construction of a secondary sewage treatment plant would cost more per household than any similar facility constructed to date in New York State. As a result of the efforts of several Village Board members, grants were obtained from state, regional, and federal programs which covered 62% of the construction costs. This was the highest percentage of financial aid received for a similar project by any village in the United States at that time (Ireson, 1972). Additionally, the Board campaigned extensively to ensure the passage of a village referendum authorizing a bond issue to pay the remaining costs. The referendum passed, the treatment plant was constructed, and resulting average sewer rates were $72 per household per year. Throughout the project, the role of the Village Board was crucial in the resolution of the problem.

The Village of Freeville (population 664), located about three miles from the Village of Dryden, provides a different example. When septic systems were checked in 1970, about half appeared to be failing. In response, the Village considered constructing a secondary sewage treatment plant. A study estimated that the cost would be $190 per household per year if no external aid were received, an amount comparable to that Dryden residents would have had to pay had no assistance been received. Outside assistance, however, did not appear to be readily available to Freeville, and the residents eventually rationalized that they were not contributing significantly to pollution of Fall Creek. The idea of a sewage treatment plant was abandoned (Ireson, 1972) and the problem of the septic tanks continues.

An organization solution that might have been considered was to join the Village of Freeville (population 664) with its downstream neighbor Etna (population 764), yielding a population nearly equal to that of the Village of Dryden (1,490), and to construct a jointly operated sewage treatment plant. Such cooperation was not considered, however, and it was impossible for Freeville by itself to qualify for external financial assistance. This illustrates the way in which social organization and economic and political factors influence decision processes and water quality issues.

Local Groups in the Three Watersheds

To better understand the roles played by local groups in resolving water quality problems within watersheds, several general social groups were identified and studied. Of particular interest were their attitudes concerning water pollution and their willingness to act in support of improvements in water quality.

Farmers

The farm households in the survey (a total of 93 in the three watersheds) engaged primarily in dairying. The average acreage was about 225. The majority of the farmers owned at least 50 animals (cattle, horses, pigs, or sheep) and consequently had to dispose of manure. Ninety-five percent of the farmers with animals were spreading manure on their fields without prior treatment; 75% spread manure throughout the year; and 10% spread in the spring and fall only. Finally, 25% stated that manure spread during the winter was not plowed under.

Farmers expressed a general willingness to support the control of water pollution. However, when specific control measures were considered which directly affected them, acceptance was far less apparent. As an example, farmers were sampled to determine their views on the spreading of manure in the winter. They were asked to consider a hypothetical referendum prohibiting the spreading of manure in winter to preserve water quality. Seventy-one percent of the farmers said they would not vote for such a proposal. Fifty percent of those who were not involved with manure disposal (38% of the farm sample) also said they would oppose the measure, in spite of the threat to water quality. The implications to farmers of spreading or storing manure are discussed in Chapter 5.

Few of the farm residents raised poultry; therefore, it can be assumed that poultry products were purchased. The question was asked of farmers: "Suppose poultry farms could significantly reduce pollution caused by manure spreading through employing improved storage practices. How much more would you be willing to pay for a dozen eggs if it allowed them to adopt these practices?" Thirty-eight percent of the farmers, compared to 23% of non-farmers, were unwilling to pay more.

The farmers' stated attitudes with regard to water quality appeared to be inconsistent with behavior. When farm residents were polled on general issues of water pollution abatement, they responded favorably; but when faced with personal sacrifice or cost, response was negative. However, this inconsistency was not limited to the farm population. In general, support was lowest for abatement measures which placed the burden of cost on a particular group, as in the case where farmers were asked to support measures to control manure spreading practices without being offered a feasible alternative. Support was greatest for measures, such as increasing food prices, which spread the costs equally among all members of society.

Water Recreationists

A comparison was made between persons who engaged in water recreation more than once a week and those who did so less than once a month. As a group, those who participated more frequently in water recreation were younger, had more children, and had higher levels of education and income.

Respondents who frequently engaged in water recreation were more likely to say that water quality had deteriorated in recent years, and noticed more signs of apparent pollution. In spite of this, they were less likely than the other group to believe that the water was unsafe for swimming, drinking or fishing. This result was unexpected; it was anticipated that the more the persons were aware of water quality degradation, the more precautions they would take to protect their health. This apparent paradox may be explained by their personal experience which had demonstrated that not all visible signs of pollution were necessarily hazardous to health. The desire of individuals to use the water for recreation may cause them to rationalize to reduce this inconsistency. For example, if a fisherman strongly desires to go fishing, he defines the stream or lake as suitable for that purpose.

Respondents who frequently engaged in water recreation differed very little from nonparticipants in their willingness to act or to pay for improvements in water quality. Since the former would

Figure 7.6. Canoeing, a popular recreational use of Fall Creek. (Photo by G. Reetz)

be expected to have the greater personal interest in preserving water quality, this result was also unexpected. In an effort to better understand this result, water recreationists were divided into groups according to the recreational activity they engaged in most frequently, such as swimming, boating, fishing, or picnicking. As expected, many people engaged in more than one activity. To eliminate overlapping, respondents were divided into groups on the basis of frequent participation in one activity and infrequent participation in the other three. When these "pure" groups were compared, no real differences emerged.

These findings support the conclusions of Field and O'Leary (1973) who found insignificant relationships between water recreation activities and social variables such as age, education, and income. Social variables, they suggest, are useful in distinguishing participants from nonparticipants, but not in predicting the choice of recreational activity among participants.

In summary, individuals who frequently participate in water recreation are more aware of water pollution than those who do not, but do not believe that it need interfere with their recreational activities. As a group, they are not more willing than infrequent participants to pay for or act in support of improvements in water quality. As individuals, they are as uncertain about their options as are nonrecreationists.

Users of Nonphosphate Detergents

Most individual efforts to improve the environment, such as recycling bottles, saving newspapers, and reducing use of fuel, are voluntary. Attempts to understand such voluntary action must consider who participates, why they do, and how much cooperation is obtained.

Data from the Owasco Lake Study were analyzed for this purpose by assessing the voluntary use of nonphosphate detergents (Gore and Gilles, 1974). The study was concerned with attitudes about phosphates as water pollutants, but did not attempt to assess the actual effects of phosphates on water quality. It was assumed, however, that the "phosphate issue" was generally known, as New York's law banning phosphate in household laundry detergents was about to come into force at the time of the study.

Persons included in the survey were asked two questions:

1. Are you using a low-phosphate detergent?
2. What is the brand name of the laundry detergent you use most often?

Of 536 respondents to the first question, 38% said that they did not use low-phosphate detergents, 43% said they did, and the remainder did not know. Only 380 respondents could name the brand of detergent they used. The detergents names were then ranked according to phosphate content using a list published by Consumer Reports (September 1970). The above responses were then checked to verify the claims made as to whether respondents actually were using brands with high- or low-phosphate content. Results showed that only 21% of those claiming to use low-phosphate detergents actually did, while 48% correctly claimed to use detergents high in phosphate. Another illustration of difference between intent and reality was that 27% of those who reported they thought they were using low-phosphate detergents were in fact using brands with high-phosphate content.

Distinctions between the different groups were sought, and it was found that those using brands with low-phosphate content were more educated and more exposed to educational mass media (television, radio and printed matter). This supports the opinion that pollution is an issue for the middle class who are better informed and educated (Morrison *et al.*, 1972).

Participants in Local Decision Making

In this study, participation was defined as attendance at public meetings or conversations with political officials. Rates of individual participation were compared with preferences for public services. It was hoped that these comparisons would reveal strategies for encouraging local policies to preserve water quality.

Respondents were asked to name the three most vital public services out of a list of ten which included sewage disposal and water supply. Respondents were also assigned a political participation score.

A statistically significant association was found between an individual's rate of participation and his opinion of the importance of sewage and water services (Gilles, 1974). Groups not normally involved in local affairs are less likely to give high priority to water quality than the small group of people usually involved in community decisions. It is possible that nonparticipants could be influenced by education campaigns, but any strategy that increases participation in public affairs may engender more opposition for programs to improve water quality than support. A village or town board may be the group most likely to support sewage treatment or water supply facilities while the general public might be less inclined, believing that community priorities lie elsewhere (Gilles, 1974; Crain, Katz and Rosenthal, 1969).

The analysis of social groups within the watersheds has shown that while concern for water quality is generally high for all groups, there are no particular groups that are outstanding in their efforts to improve water quality. Neither, however, is there any organized opposition to such efforts. Maximum support for improvements in water quality may be received when costs are born equally by all members of society, when people are informed about politics, and when that information is clear and specific. Local government appears to be the best choice for providing leadership in improving water quality.

IMPLICATIONS FOR INTERPRETING RESEARCH RESULTS

Although policy formulation is generally a public affair and thus is subject to political constraint, the scientific investigator is under an obligation to initiate dialogue between concerned groups. One means of fulfilling this obligation is for the investigator to interpret his results so they are relevant to the planning and decision-making processes. A study such as the one reported here is an opportunity to increase communication between those interested in resource policy.

Results of perception studies particularly need careful interpretation. Often such studies reveal important inconsistencies between an individual's perception of reality and the same individual's action. For example, a person may feel a stream is polluted and still swim in it. He may believe in the effectiveness of low-phosphate detergents but not use them. The first example is one of what has been called "cognitive dissonance" (Festinger, 1956). This causes the individual to argue "extenuating circumstances" which make actions consistent with belief. The second example is a special example of cognitive dissonance known as the "tragedy of the commons" (Hardin, 1968) where an individual realizes that his action is futile unless everyone acts as he does. In this case, a person may not make the effort to conform to his beliefs because he does not believe (and rightfully so) his action will make a difference.

The "contradictions" due to cognitive dissonance found in preliminary reports of this study have led public officials and extension workers to request more interpretation of findings.

SUGGESTIONS FOR FURTHER RESEARCH

The study of attitudes and perceptions in relation to the environment remains formative. Advantages and limitations of such research for policy formation have yet to be determined, and important questions remain unanswered:

—What is the useful life span of attitudinal data on the environment?
—How quickly does concern for the environment change?
—How responsive is the general public to persuasion by education and dissemination of information?
—How is concern for the environment changed, for example, by food and energy crises?

There is a need for creative approaches to the study of environmental decisions at the community level. Ireson (1972) argues that decisions are influenced by social structural factors, such as institutional resources as well as by individual attitudes. Although this study was limited primarily to an investigation of individual attitudes and behavior, future efforts should also consider other factors. Cornell's Department of Rural Sociology is assessing the usefulness of combining perception and attitudinal studies with socioeconomic, organizational, and institutional variables in the context of environmental control in its current study of the relation of social structure and environmental issues in the Hudson River Basin.

SUMMARY AND CONCLUSIONS

While the findings in the three watersheds were specific to the populations studied in central New York, implications are believed to be relevant to other states and regions.

While technically the interdisciplinary research found Fall Creek to be free from serious loadings of either nitrogen or phosphorus (see Chapter 3), the residents perceived a serious and worsening situation of pollution. Several things contributed to this anomaly. First the environmental movement was gaining momentum in 1971. Some visible evidence of pollution such as scum on the water, slime on the rocks, trash in the stream was reported by 70% of the residents. The County Public Health Department closed Fall Creek for swimming in this period. Many concluded that the water was biologically unsafe. The real reason had more to do with the fact that lifeguards were difficult to obtain and there was a lack of liability insurance.

Elsewhere in New York State heightened public awareness to water pollution was reflected in the $1.5 billion Environmental Bond Act of 1972 and a Harris Poll which identified pollution and especially water pollution as key issues of concern (Harris, 1971).

It is clear the public has become more sensitized to environmental issues. The mobilization of the environmental movement, public debate, and mass media have created a public awareness of problems pertaining to the environment. However, the appropriate action to be taken, by whom and at what cost or benefit depends upon which issue, and which standard.

The responsibilities of individuals concerning nutrient management and control were identified as they relate to the environment. For example, farmers have responsibility relative to the uses of fertilizers, pesticides, animal control, barnyard runoff control and the spreading and management of manure including odor control. Residents have responsibilities of trash disposal, management of runoff and seepage from heavily fertilized lawns and gardens, maintenance and control of septic tanks and disposal of waste-water from laundry and kitchen facilities. All individuals who utilize the stream for recreational purposes have responsibilities of refraining from throwing refuse into the stream and otherwise using it as a receptacle for leftover waste.

The mechanisms for inducing responsiveness toward management and control of nutrients and water quality are education, incentives and regulatory measures. While incentives and voluntary efforts are recognized as necessary, the principal constraint was the suspicion that others were failing to do their part. The reality became recognized that pollution and maintenance of water quality placed an unequal responsibility on different residents of the watershed. Such recognition was shown by residents who expressed willingness to pay more for farm products if increased production costs resulting from pollution controls placed heavy burdens on farmers.

Furthermore the resident populations of the watersheds felt the existing pollution control measures were ineffective, the standards too lenient and that major efforts should be expended to address the problems. There was expressed opinion that sewage effluent was more of a problem for the streams and lakes than farm runoff from fertilizers or insecticides.

Generally residents expressed concern about the value of their own individual compliance with environmental controls unless others also conformed. Rather than voluntary compliance, greater confidence was reported in uniform regulatory measures applied equally to everyone. The preferred level of regulation was local government. However, while county government contains basic regulatory agencies such as the Department of Health and Office of Planning, still the residents were aware that streams flow across several political boundaries like municipalities, townships and counties. Several recognized that, ideally, water pollution control agencies should incorporate whole watersheds. Another approach would be for multicounty regional government bodies or for the state to assume regulatory responsibilities.

A number of special insights have been gained from the study which have more general application to lake and watershed management and control.

1. There is evidence of a propensity on the part of individuals in a watershed to employ a type of wilderness psychology which implies that each individual has a God-given right to do as he pleases

with his property, and no one has a right to interfere with his traditional habits and customs of use. Correspondingly, there is a tendency for people to be indifferent to what happens beyond the limits of their own property.

Such attitudes deny the existence of ecosystem interaction. This concept perhaps needs to receive greater attention on the part of environmentalists and others who are in a position to influence modes of behavior through education, incentives and regulatory means.

2. Another subtle form of group-level territorial argument has to do with the numerous boundary jurisdictions of local government units through which streams flow. The implications here are that one township or municipality has certain responsibilities for that portion of the stream situated within its physical boundary. Moreover, the inference is that different standards and degrees of enforcement of local government regulation are appropriate for different portions of the stream as it winds its way from origin to destination. This pattern of legislative management, or in some cases, the lack of management brings a haphazard pattern to any systematic form of administration or control. Since the watershed is an ecological entity as well as a sociological entity, innovative ways need to be derived to also make the watershed a homogeneous administrative entity. Arrangement such as watershed associations or multicounty regional governing bodies with properly delegated powers of administration and control are possible approaches to overseeing the complexities of the problems.

3. Traditionally many administrators think of a watershed in terms of its physical and geographical characteristics. Another frame of reference is to think of a watershed in terms of its flowing through a social, economic and political landscape. The latter may have as much uneven terrain with notable peaks and valleys in a comparative sense as does the physical and geographic landscape. Each local area of jurisdiction reflects this variation in terms of its economic, social and political composition.

4. There is strong psychological support to base management control options on what might be termed the "a-ach effect" as people contemplate the notion of drinking others' sewage waste. However, they appear less concerned about others drinking their sewage as evidenced by the reluctance of municipalities to install or upgrade sewage plants unless forced to do so by higher levels of government which also pays much of the cost.

Several instances were found in the interdisciplinary study which highlight seeming discrepancies between available knowledge (technology) and behavioral action. Consider for example the problem of inducing the household residents on Long Island to change their practices of using phosphate detergents. In 1973 a ban on detergents was adopted by the tricounty legislative bodies of Queens, Nassau, and Suffolk making it illegal in those counties to purchase and use phosphate detergents. This action reportedly generated a brisk enterprise of bootlegging detergents from surrounding counties. Apparently some residents were not sure of the cleaning efficiency of nonphosphate detergents and rationalized that "their little bit of phosphorus wouldn't make any difference."

Another example of the dichotomy between expressed attitudes and beliefs is the problem household residents face with faulty septic systems. The study showed that occupants of summer cottages which were tightly packed around the shore of Canadarago Lake had good reasons to be suspicious of septic tank leakage into the lake. Nevertheless these occupants were very reluctant to accept the cost of linking their systems with the new sewerage and tertiary treatment facilities since many of them only used their cottages for two or three summer months of the year. Many were very critical of neighbors with faulty septic systems. In response to a later question, however, many reported they had never had their own septic tanks inspected even though the systems had been installed for 20 or 30 years.

The above cases document behavioral inconsistencies. They suggest people fail to do many of the things they could or should. The explanation for this apparent discrepancy in beliefs, values and attitudes lies in the nature of the assumptions made. Any given individual would defend his or her actions as being internally consistent and rationalize action on the grounds that others are equally inactive. Perhaps more importantly, many individuals are uninformed, and do not know what the appropriate action should be nor its costs and consequences.

The findings reinforce the conclusion that response of individuals to nutrient management and control is crucial. The additive and cumulative effects of positive or negative action make a world of difference. There are definable limits, however, beyond which individuals acting by themselves have diminishing impact. The still uncharted arena of social, economic and political institution-building, however, is where environmental management and control strategies primarily fail to match the magnitude of the problems.

REFERENCES

Borton, T., and C. Warner. 1970. The Susquehanna communication-participation study. Institute for Water Resources, IWR Report 70–6. U.S. Army Corps of Engineers, Alexandria, Va. 128 pp.

Capener, H. R., J. D. Francis, P. H. Gore, and D. R. DeLuca. 1974. Perceptions of environmental quality issues in the Hudson River region. Cornell Community and Resource Development Series Bulletin 7. Cornell University, Ithaca, N.Y. 43 pp.

Citizens Advisory Committee on Environmental Quality. 1972. Citizens make the difference: case studies of environmental action. U.S. Government Printing Office, Washington, D.C. 71 pp.

Consumer Reports. September 1970. Consumers Union of the U.S., Inc., Mount Vernon, N.Y. Vol. 37(9).

Crain, Robert D., L. Elihu Katz and Donald B. Rosenthal. 1969. The politics of community conflict. Indianapolis: Bobbs-Merrill. pp. 269.

Dasgupta, S. May 1967. Attitudes of local residents toward watershed development. Social Science Research Center, Mississippi State University, State College. 30 pp.

Davies, J. D., III. 1970. The politics of pollution. Pegasus, New York. 182 pp.

Erskine, H. 1972a. The polls: pollution and its cost. Public Opinion Quarterly 36 [spring]: 120–135.

Erskine, H. 1972b. The polls: pollution and industry. Public Opinion Quarterly 36 [summer]: 263–280.

Festinger, Leon. 1956. A Theory of Cognitive Dissonance. Stanford: Stanford University Press. 261 pp.

Field, D., and J. O'Leary. 1973. Social groups as a basis for assessing participation in selected water activities. Journal of Leisure Research 5(2) [spring]: 16–25.

Finley, J., and A. Hickey. 1971. A study of water resource public decision making. Cornell University Water Resources and Marine Sciences Center Technical Report 37, Ithaca, N.Y. 38 pp.

Finley, J., and J. Baker. 1972. Social elements in environmental planning. Battelle Research Outlook 4(2):8–11.

Gilles, J. L. 1974. Non-participation and community decision-making. M.A. Thesis. Cornell University, Ithaca, N.Y. 62 pp.

Gore, P. H., and J. L. Gilles. 1974. Voluntarism in environmental action. Human Ecology Forum 4(3) [winter]:12–13.

Gore, P. H., S. Wilson, and H. R. Capener. 1975. A sociological approach to the problem of water pollution. Growth and Change 6 [January]:17–22.

Hardin, Garrett. 1968. The tragedy of the commons. Science 1962 [September]: 1243–248.

Harris, Louis, and Associates. May 1971. The public's view of environmental problems in the state of New York. Study 2119. 136 pp.

Haskell, E., and V. Price. 1973. State environmental management: case studies on nine states. Praeger Publishers, New York. 283 pp.

Heberlein, T. A. 1974. The three fixes: technological, cognitive, and structural. In D. Field, J. C. Bureen, and B. F. Long, eds. Water and community development: social and economic perspectives. Ann Arbor Science Publishers, Ann Arbor, Michigan. 386 pp.

Hendee, J. C., R. P. Gale, and J. Harry. 1973. Conservation, politics and democracy. Journal of Soil and Water Conservation, 24[November/December]:212–215. Quoted in S. L. Albrecht. 1973. Environmental social movements and countermovements. In R. R. Evans, ed. Social Movements. Rand McNally College Publishing Co., Chicago.

Hetling, L. J. 1971. Sources of nutrients in Canadarago Lake. New York State Department of Environmental Conservation Technical Paper 3.

Holcomb, B. 1974. Environmental quality and leadership in northern New Jersey: an exploratory investigation. Department of Geography, Livingston College, Rutgers University, New Brunswick, N.J. 72 pp.

Ireson, W. R. 1972. Community decision making as collective response: three New York villages. M.A. Thesis. Cornell University, Ithaca, N.Y. 82 pp.

Lionberger, H. F. 1960. Adoption of new ideas and practices. Iowa State University Press, Ames.

Meadows, D. H., D. L. Meadows, J. Randers, and W. W. Behrens III. 1972. The limits to growth: a report for the Club of Rome's project on the predicament of mankind. Potomac Associates, New York. 207 pp.

Morrison, D., K. Hornback, and K. Warner. 1972. The environmental movement: some preliminary observations and predictions. In W. Burch (ed.). Social behavior, natural resources and the environment. Harper and Row, New York. 374 pp.

Murch, A. W. 1971. Public concern for environmental pollution. Public Opinion Quarterly 35 [spring]:100–106.

Oglesby, R. T., L. S. Hamilton, E. L. Mills, and P. Willing. 1973. Owasco Lake and its watershed. Cornell University Resources and Marine Sciences Center Technical Report 70. Ithaca, N.Y. 52 pp. and appendices.

Peterson, J. H. 1971. Community organization and rural water system development. Water Resources Research Institute, Mississippi State University, State College. 53 pp.

Ridgeway, J. 1971. The politics of sociology. E. P. Dutton and Co., New York. 222 pp.

Rogers, E. T. 1962. Diffusion of Innovations. The Free Press, Glencoe. 367 pp.

Saint, W. S., H. R. Capener, P. H. Gore, and D. R. DeLuca. 1974. Canadarago lakeside residents: perceptions on safeguarding water quality and maintaining adequate septic disposal systems. Cornell Community and Resource Development Series Bulletin 8. Cornell University, Ithaca, N.Y. 12 pp.

Simon, R. J. 1971. Public attitudes toward population and pollution. Public Opinion Quarterly 35 [spring]:95–99.

Smith, G. J. C., H. J. Steck, and G. Surette. 1974. Our ecological crisis: biological, economic and political dimensions. Macmillan Publishing Co., New York. 198 pp.

Thompson, D. L. (ed.). 1972. Politics, policy and natural resources. The Free Press, New York. 452 pp.

Utah Water Research Laboraory. 1971. Water resources planning and social goals: conceptualization toward a new methodology. Publication PRWG-94–1, Utah State University, Logan, Utah. 89 pp.

Wilkinson, K. P. July 1966. Local action and acceptance of watershed development. Water Resources Research Institute, Mississippi State University, State College. 43 pp.

8

Assessment and Management of Nitrogen and Phosphorus

Chapter 8

ASSESSMENT AND MANAGEMENT OF
NITROGEN AND PHOSPHORUS

The aim of this chapter is to integrate, in terms of assessment and management, the uses and fate of nitrogen and phosphorus in the human food web. In the case of phosphorus, the studies previously described were undertaken in central New York. However, many of the results have application beyond this region and these will be discussed in this chapter. The nitrogen studies were less regional in character and general issues have already been discussed. This chapter will continue the discussion.

A reader who has followed the account set forth in previous chapters will have become aware of several threads in the arguments. It is an aim of this chapter to synthesize some of the seemingly disparate issues. Some chapters have discussed mainly nitrogen and others phosphorus. The separate treatment results partly from the very different characteristics in behavior of nitrogen and phosphorus, and partly from varying degrees of seriousness of the problem posed by the two substances.

The mobility of inorganic nitrogen in soil is usually high; the inorganic forms are labile and the atmosphere serves as a source and sink. On the other hand, the inorganic forms of phosphorus are relatively immobile in soil, and there is no gaseous phase in its cycle.

In addition to the rather different chemical properties of the two elements, the influence of each element on water quality varies considerably among regions. In northeastern areas of the United States, the concentration of nitrate-nitrogen in municipal water supplies rarely exceeds 10 mg/l except on Long Island and perhaps in other fairly restricted local areas. In the Northeast there are few large areas of farming with very high use of nitrogenous fertilizers, few areas where many septic tanks recharge the ground water used as municipal water supplies, and there are large amounts of recharge from precipitation (30 to 50 cm/yr, or 3 to 5×10^6 liters/ha/yr) which dilutes any nitrate added. This is not true of many other areas in the United States. For example, in Illinois the levels of nitrate-nitrogen in several streams used as water supplies are near 10 ppm for portions of the year (Harmeson and Larson, 1969), and in California concentrations in recharge water often exceed this standard (Lund, *et al.*, 1974).

With respect to the enrichment of surface waters, some reports indicate that nitrogen is a limiting nutrient in highly enriched waters where phosphorus is abundant. However, in the Northeast, algal growth in lakes appears to be primarily limited by available phosphorus. Thus, phosphorus control in these lakes may be a matter of high priority. Probably this is also true of some other surface waters in the United States. However, the enrichment of rivers by phosphorus is of less concern for several reasons. Phosphorus does not accumulate in the water of rapidly-flowing rivers. Furthermore, movement of the water scours the river bed, especially during high flow, and the turbidity of such rivers reduces the availability of sunlight for growth of aquatic plants. An extreme example would be the Mississippi and its tributaries. Probably phosphorus control has low priority in such streams but may be vitally important in subwatersheds from which drainage water feeds lakes and reservoirs.

In this chapter, the conclusions of general applicability will be discussed based on the outlines of the flows of nitrogen and phosphorus reconstructed from previous chapters. General economic issues are considered where relevant to the general management of nutrients in rural watersheds. Additionally, some implications of the water pollution control act amendments of 1972 (PL92–500) will be briefly referred to in the following discussion. A provision of the act that is highly relevant to these studies is that it requires the development by appropriate agencies of information including "(1) guidelines for identifying and evaluating the nature and extent of non-point sources of pollutants, and (2) processes, procedures, and methods to control pollution resulting from . . . agricultural and silvicultural activities, including runoff from fields and crop and forest lands. . . ." These two clauses represent, respectively, guidelines for scientific investigation (assessment) and for management. This division, assessment and management, will be applied in the following discussion of the flows of nitrogen and phosphorus.

In previous chapters, major aspects of the sequence of effects and their causes in the nutrient flows were retraced from the lake to the animal production unit. It is now convenient to consider the flow in reverse, starting with human and animal wastes as an initial source of the substances in the watershed.

HUMAN AND ANIMAL WASTE

Assessment

One very clear source of phosphorus input to lakes is municipal sewage treatment plants (mostly primary and secondary in New York) which discharge effluent directly into lakes. The phosphorus content of domestic sewage ranges from 1 to 2 kg/cap/yr, depending primarily on whether detergents containing phosphates are used. In a secondary treatment plant, approximately 20% of the phosphorus in the influent is removed in the sludge during treatment. With tertiary treatment for phosphorus removal, 75 to 90% of the influent phosphorus is incorporated into sludge.

Experience with one secondary sewage treatment plant in the Fall Creek watershed suggests that a substantial fraction of the soluble phosphorus discharged from the plant directly into the stream was transformed into relatively insoluble particulate forms as the water flowed downstream, and thereby it was rendered biologically unavailable. If this is a general occurrence, then the impact of municipal sewage treatment plants decreases as distance from the lake increases.

The estimates of phosphorus contributed by nonsewered households are subject to considerable doubt. In Chapter 2 the estimate was 50% of the phosphorus content of domestic sewages, while in Chapter 3 the estimate was 10%. The estimate in Chapter 2 reflects measurements in one stream and the following intuitive weighting factors: a) populated areas tend to be concentrated near streams and lakes and hence transmission losses are minimum and biological impact is maximum; b) these settlements are fairly old and many of the septic tank disposal fields overflow into storm drains; and c) areas for location of these disposal fields are limited in size. On the other hand, the estimate in Chapter 3 is based on measurements of dispersed housing units in rural areas where transmission losses are high and where adequate area is more often available for disposal fields. Regardless of which estimate is chosen, septic tank disposal fields appear to "leak" less phosphorus to lakes than municipal sewage collection and disposal systems unless such systems include tertiary treatment for phosphorus removal. If the disposal fields for septic tanks are in well-drained soils and these systems are properly maintained, then the loss of phosphorus to lakes and streams is probably near zero. Hence well-designed and properly-functioning home sewage disposal systems are roughly comparable to municipal systems employing tertiary treatment for removal of phosphorus. Even when home disposal systems are not well-designed and maintained, they appear to put less phosphorus in the streams and lakes than municipal treatment plants with secondary treatment.

The nitrogen content of the human diet is about 6 kg N/cap/yr, and is the main source of nitrogen in domestic sewage. During treatment the nitrogen may be removed from the wastewater by being volatilized as NH_3, denitrified, discharged in the effluent or retained in the sludge. The proportion of each varies with design and operation of plants, and generalizations are not easily made. However, in a well-designed and operated secondary treatment plant intended to provide nitrification most of the nitrogen in the influent will be discharged as NO_3 in the effluent. When nonnitrified effluent is discharged to streams and the dilution afforded by the river is low, NH_3 content may pose some hazards because of its toxicity and also because it is oxidized to NO_3. In either case, inorganic nitrogen in the effluent may elevate levels of NO_3 in the stream to levels which exceed the public health standard for nitrate. Nitrogen from septic tank disposal fields may also be converted to NO_3 and leached to ground and surface water where it may constitute a similar potential problem.

The amount of nitrogen and phosphorus which is lost to water from animal production units is uncertain. Where units are small and scattered throughout a watershed, it may be impractical to monitor their impact on the environment although it may be significant locally. In the case of large-scale units with high-density animal confinement the monitoring is easier, but the effects of run-off and seepage on ground and surface waters in specific cases can only be determined by direct measurements.

Management

As indicated in Chapter 1, the use of phosphate detergents has been prohibited in Canada and in several regions in the United States including New York State. Although the effect of this prohibition has yet to be fully determined, it is believed that at least in

New York State the phosphorus in domestic sewage has been reduced by about 50% (Hopson, 1975). Even if allowance is made for the poorer efficiency of the detergents and effects on washing machines and clothes, this reduction in phosphorus probably was achieved more cheaply than any other single foreseeable measure including tertiary sewage treatment. At the cost of having to wear shirts with less luster rather than with the dazzling whiteness lauded in advertisements for detergents, the enrichment of surface waters has been reduced. It may, however, be admitted that the economic balance between sartorial disutility and higher transparency in lake water possibly rests on what is intrinsically incommensurable.

Both phosphorus and nitrogen can be removed in waste treatment. As indicated in Chapter 5 (Table 5.17), the estimated cost of removing phosphorus by tertiary treatment was $12 per kg of phosphorus removed in a small sewage works and less in larger plants. This cost was substantially less than that estimated for other sources of phosphorus with the possible exception of barnyard runoff. However, practical problems must be admitted. Although removal of phosphorus may be relatively the least expensive, the total cost of providing the necessary sewerage and sewage treatment can still be prohibitive for small communities as suggested in Chapter 7. From the work discussed there, it would appear to be impossible for these communities to meet the costs of installing either secondary or tertiary treatment systems without substantial state or federal assistance. The community effort necessary to put into effect such schemes may call for considerable coordination between the government agencies involved, as well as strong leadership from committed individuals.

Tertiary treatment and the ban on phosphate in laundry detergent are alternative methods of reducing phosphorus in wastewater. The ban on phosphate detergents was relatively easy to implement, apparently of low cost, and had a substantial impact on the input of phosphorus to lakes in central New York. As tertiary treatment is added to more and more sewage plants, perhaps the ban on phosphate detergents will become less necessary. Some residents of sewer districts that have installed tertiary treatment argue that they should be allowed to purchase phosphate detergents because they have committed themselves to the cost of phosphorus removal by tertiary treatment. Implementation of such a differential policy would be difficult. As more communities install tertiary treatment, there may be increased pressure to repeal the phosphate ban. However, there may be substantial savings in chemical costs of tertiary treatment if the phosphorus content of sewage has been previously reduced by a phosphate detergent ban.

The phosphorus removed from wastewater during treatment is incorporated into sludge, the disposal of which may in itself pose formidable problems for certain localities. On the other hand, nitrogen is usually removed by reducing nitrogenous waste to a gaseous form, such as ammonia or nitrogen gas, which is released to the atmosphere, and does not increase the amount of sludge produced during treatment.

Both substances may similarly be removed from animal wastes. In the case of nitrogen, the most likely methods to be adopted are ammonia stripping and the nitrification-denitrification process (Chapter 6). Ammonia stripping is a technical possibility, but requires a degree of pH control unlikely to be continuously achieved by those operating on the farm animal waste treatment plants. As discussed previously, some uncontrolled ammonia losses will inevitably occur depending on aeration or agitation of the wastes, time of storage and handling, pH and methods of disposal.

Losses of nitrogen in ammoniacal form are also inevitable in any system that accumulates and dries manure such as the high-rise system of raising poultry. In this system, manure is allowed to collect in the form of cones under the cages. Fans circulate air over the manure and reduce the moisture content by almost 50%. Although losses of ammonia are continuous, there is an absence of offensive odor in well-managed aerated systems. The manure is usually removed annually and disposed on land. Nevertheless, there is an increasing difficulty for poultry farmers, such as those in the Catskill recreational area of southeastern New York, to find sufficient land on which to dispose of the manure without creating a potential environmental hazard. Most of the expansion of poultry production in the state in the last few years has taken place in upstate New York, where land is available and can be integrated into a crop-animal production system from the outset. The economic and legal factors underlying this trend may be such that waste management, animal production and crop production must be envisaged as a single entity in farming operations in the future. Liquid systems of manure collection or storage that are anaerobic will tend to conserve nitrogen within the manure but these systems result in extremely offensive odors at spreading time which are not tolerated by neighbors.

With respect to phosphorus, it was argued in Chapter 6 that the removal of phosphorus in animal wastes was not economical due to the nature of the wastes, and even impractical given current methods of animal production. Given that animal production facilities are usually in the vicinity of grass, crop, or brush land, it would appear that disposal on these areas is the obvious strategy. With proper land and crop management, most of the phosphorus in the animal wastes applied to the land would be retained at the place of application.

Of the various agricultural sources of phosphorus, one of the easiest and least expensive to control appears to be barnyard runoff. Both initial investment and the annual operating costs of barnyard runoff control systems are relatively low, at least in comparison to manure storage, a conclusion confirmed by others (Good et al., 1973; Buxton and Ziegler, 1974). However, the amount of phosphorus that could be reduced by barnyard runoff control is uncertain and available data indicate that it varies widely according to the nature of barnyards. There is a need to increase the level of knowledge about phosphorus losses from barnyards and feedlots.

It is believed that barnyard runoff control or manure storage would have a greater adverse economic impact on small dairy farms (20 to 50 cows) than on larger dairy farms. Larger farms tend to have lower production costs because of more efficient utilization of labor, equipment and buildings. Unfortunately, the cost of pollution control generally is higher per unit of production on the smaller farms, as demonstrated by data for both barnyard runoff control and manure storage (Chapter 5). Consequently operators of smaller dairy farms are under pressure to either discontinue production or to enlarge their operations not only for greater efficiency but also to reduce the unit cost of pollution control. These conclusions have been confirmed by others (Buxton and Ziegler, 1974; Good et al., 1973). It follows that regulations which require installation of pollution control facilities will increase economic pressure on operators of smaller dairy farms. A similar situation exists for other livestock (Van Arsdall et al., 1974) and poultry farms.

The economic and legal factors already referred to are such that there is a nationwide trend toward more complete treatment and disposal of human wastes and control of animal wastes to protect the environment, complemented by the aim to conserve resources. Both objectives are at least implicit in recent legislation. Regulations which encourage the integration of waste treatment with recycling the wastes should be encouraged. Thus, as described in the example of the poultry farmer, there could be a general shift from viewing manure handling as waste disposal to an opportunity for recycling nutrients.

FLOWS IN CROP PRODUCTION AND ON LAND

Assessment

In Chapter 4 the nitrogen flows in the human food web were examined for the nation as a whole and for the Fall Creek watershed. In both cases large amounts of nitrogen were unaccounted for; presumably the unaccounted-for nitrogen is lost mainly as nitrate in ground water, NH_3 volatilization to the atmosphere and as N_2 via denitrification. In previous chapters NH_3 volatilization and nitrate in water were identified as major unaccounted-for forms of nitrogen which may have environmental consequences. For example, NH_3 volatilization from manure was estimated as 4 million tons of nitrogen annually in the United States. The NH_3 volatilization to the atmosphere may have an influence on the atmosphere *per se*. In addition, the atmosphere serves as a vehicle for dispersion of NH_3, and its reaction products, over large areas before their redeposition in the landscape.

The significance of these losses of ammonia to the atmosphere has yet to be fully established. According to Hidy (1973), a major effect of ammonia in the atmosphere is to promote the conversion of SO_3^{2-} to SO_4^{2-} in aerosols and to neutralize sulfuric acid. On the other hand, McConnell (1973) postulates several atmospheric reactions for ammonia that could lead either to formation of large amounts of nitrous oxide or destruction of large amounts of the same, depending upon the importance of several alternative reactions.

In the central New York area, Likens (1972) reported an average input in precipitation of about 8 kg N/ha/yr, approximately equally divided between nitrogen in ammoniacal and nitrate forms. If the ammoniacal nitrogen were derived from gaseous ammonia, then it has neutralized an amount of acidity approximately equal to one-fourth of the acidity remaining in the precipitation, which in the northeastern United States is now pH 4.0, or about 10^{-4} M H^+/l. Thus with respect to acidity in precipitation and removal of SO_2, ammonia has a beneficial effect. On the other hand, if the nitrate forms were derived from oxidation of ammonia, these reactions could be a source of a portion of the acidity now in the precipitation, according to some of the hypothesized reactions described by McConnell (1973).

With respect to redeposition of fixed forms of nitrogen (NH_3, NO_3^-, etc.), the total input from atmospheric sources is largely unknown, but consists of the sum of nitrogen carried in the precipitation, fallout of particulate matter, and gaseous absorption. The latter two components are largely unmeasured. In addition, the landscape is likely to consist of a mosaic of sources and sinks, i.e., animal production units and manured fields dispersed among non-agricultural sinks such as acid soils, vegetations, etc. The degree of dispersal of ammonia from the various sources will vary widely with local climatic conditions and season.

Deposition of manure and the accompanying volatilization of ammonia are not limited to domestic animals. One can speculate that before the western range became the "home" of the cowboy, the skies "that were not cloudy all day" received a large daily quota of ammonia from the manure from the massive herds of buffalo.

The models described in Chapters 4 and 5 attempted to represent, in varying detail, the movement of nitrogen and phosphorus on and through the soil. There are reservations regarding these and other such models which must utilize sparse data and concepts with admitted shortcomings. Unfortunately, in the absence of other scientific means of obtaining answers to the questions dealt with by the models, they must suffice. It is relatively easy to dispose of the models with criticism, but difficult to propose better solutions. However, the results they provide may be interpreted more intelligently if the difficulties are made explicit.

The phosphorus models are based on the universal soil loss equation, a phosphorus enrichment ratio, a delivery ratio and, in some cases, the percentage of phosphorus that is estimated to be biologically available. The underlying assumption of most current efforts to model the movement of phosphorus on land is that it is geochemically an immobile substance. Phosphorus reacts very strongly with the soil. Eroded soil material which is transported by overland flows of water, therefore carries phosphorus. Estimation of this phosphorus loss requires estimation of the phosphorus content of eroded soil particles. Such estimation is complicated by the fact that very small particles are most erodible. These particles have a higher relative reactivity with phosphorus and hence, compared to the soil as a whole, are proportionately richer in phosphorus.

Allowance is made for this expected phenomenon by the enrichment ratio.

Further complication arises because soil loss models were originally developed from small plot studies. When they are extended to whole watersheds, it is necessary to allow for the redeposition of some sediment as it is transported across the landscape. On a watershed scale, the proportion estimated to leave the stream is estimated by the delivery ratio, which is the quotient of observed loads in streams to the total soil loss predicted by soil loss equations for the entire watershed. Thus the sequence of the estimation is (1) the total soil loss or erosion, (2) the enrichment ratio, and (3) the delivery ratio.

Large errors may be incurred in such a series of calculations. For example, so many variables govern (1) and (2), it is improbable that either will be estimated with reliability. With respect to (3), the load estimated in the stream, itself likely to be in error, has many sources other than soil erosion from cropland, such as the banks and bed of the stream.

Finally, the above procedure only estimates the loss of total phosphorus from land rather than that phosphorus which is biologically available. If attention is directed to soluble phosphorus, soil erosion may recede in importance whereas barnyard runoff, sewage works, and even tile drains and ground water seepage may become of primary significance. Models that estimate total phosphorus rather than biologically available phosphorus (BAP) are likely to be misleading for management of phosphorus in lakes.

Despite all of the foregoing, it may be argued that although current models of the loss of phosphorus from land are in absolute error, they may still provide insight into the relative effects of various control measures, such as those outlined in Chapter 5. No other comprehensive assessment of such measures appears to be available.

There is abundant scope and need for new research. This could take the direction of attempting to account for phosphorus movement in terms of the total drainage system, working from first principles of geochemistry and hydrology. This strategy could, admittedly, further complicate an already difficult research task, and may be less practicable given its greater requirements for data.

Given the solubility and relative mobility of nitrate-nitrogen in soils, it may be easier to construct computer models representing the broad flows of nitrogen than those of phosphorus, assuming that the movement of ground water can be adequately modeled. Such a model, constructed as part of these investigations, is nearing completion as this is being written (Tseng, 1975).

Management

Nutrient flows on land may be managed by timing the application of fertilizers and manure, by controlling the amounts applied, and through cropping and soil conservation practices. Nitrogen on the soil surface and liable to removal in runoff may be relatively unimportant compared to inorganic nitrogen in ground waters. The reverse is usually true for phosphorus in that removal by surface runoff will probably be greater than losses from leaching. For example, runoff water from fields with surface applications of manure usually contains high concentrations of phosphorus. However, such runoff occurs largely on the more poorly drained soils and when rain falls or snow melts on frozen soil (Chapters 3 and 5; Klausner et al., 1975).

Storage of manure to avoid runoff of phosphorus from winter spreading of dairy manure in northern climates requires relatively large investments and annual costs. In addition, removal of manure from storage in the spring competes for the available labor and equipment required for spring planting (Casler and La Due, 1972; Good et al., 1973; Ashraf and Christensen, 1974). Application of manure in the spring may increase its value for crop production, but the evidence is unclear and controversial. Also, it is not known how much such a procedure would reduce biologically available phosphorus flowing to surface waters. Measurements made at the edge of plots indicate that losses following the spreading of manure in winter may be higher than those in spring, especially if the manure is applied to frozen ground and rain-induced runoff follows. However, in other cases, winter spreading may have lower losses and it may be that winter spreading of manure is less a source of phosphorus in surface water than is generally assumed. Virtually no data are available which relate winter spreading of manure to phosphorus losses to streams and lakes. If winter spreading of manure was shown to produce unacceptable losses of phosphorus to lakes, development of cheaper methods of storage would encourage farmers to avoid winter spreading. One should remember that liquid storage of dairy cattle manure under anaerobic conditions will produce offensive odors which will drift considerable distances when the manure is removed from storage.

The models reviewed in Chapters 4 and 5 also considered ways of altering production practices to reduce losses of nutrients from land. Under certain conditions it was found that such controls could be achieved with minimal decrease in crop output or net income. In some cases, however, to meet controls it was necessary to change to non-row crops with lower soil and nutrient losses, resulting in substantial decrease in net income of farmers.

The research reported in Chapter 5 indicates that the cost to farmers in terms of percentage reduction in net income or the cost per kg of BAP prevented from entering the lake is high, except for the reduction of the first 10 to 20% of BAP. In addition, the estimated total reduction in BAP losses that could be achieved by reducing soil erosion is relatively small compared to the reduction that could be achieved by tertiary treatment of municipal sewage.

Concerning nitrate nitrogen, levels in Fall Creek were always well below 10 mg/l. These levels would probably increase if a larger proportion of the watershed was planted to corn. For example, it was estimated (Chapter 3) that if 40% of the watershed were used for corn production, compared to the present 15%, then the standard of 10 mg NO_3–N/l might be exceeded in Fall Creek, if current production practices were contained. Such a situation could be alleviated to some extent by altering the rate and timing of application of manure and nitrogen fertilizer. An example of how yields could be maintained while using less fertilizer was shown by the investigation of fertilizer use on potato crops on Long Island.

Two other conclusions drawn from the studies are relevant to the national problem of managing nutrients. One is the possibility that control of adverse environmental effects of agriculture in one watershed or area could have environmental effects in other watersheds or areas. The studies described in Chapter 5 showed that reduction of phosphorus losses, by changing from row to hay crops, led to increased purchase of feed from outside the watershed. To the extent that production of this feed in some other watershed or area produces nutrient losses, such losses are merely transferred. Similarly, under some of the restrictions imposed, milk production in the watershed was reduced. To the extent that consumers want and are willing to pay for milk, it will be produced elsewhere, again with the possibility of transferring nutrient losses to other watersheds. This has implications for units of government which may be able to reduce nutrient pollution of water from agriculture, but only at the expense of increasing nutrient losses in other areas with less stringent controls. Additionally, if all units of government instituted similar controls, thereby reducing or eliminating the shift of nutrient losses from one area to another, food production may be reduced and costs of food to consumers increased. This has been demonstrated by other research projects, particularly with respect to reductions in the level of nitrogen fertilization (Taylor and Swanson, 1975).

Second, with respect to trade-offs among environmental problems, it appears that actions taken to reduce one type of adverse environmental effect may produce other adverse environmental effects. For example, manure storage intended to reduce runoff losses of nutrients from manure, usually will produce very unpleasant odors at the time the storage is emptied. In this case, while runoff losses may be reduced (although even this is uncertain), another environmental problem is increased. Many other environmental trade-offs occur and should be considered in environmental research and regulations.

The studies estimating the loss of phosphorus from various sources to lakes clearly identify the lack of knowledge concerning the relations between watershed activities and phosphorus losses. This suggests that data collection programs and analyses are badly needed to increase knowledge about relations between watershed activities, phosphorus losses and water quality. The results of the above studies should be interpreted qualitatively, rather than quantitatively, and great care is needed in interpreting and applying the results or the techniques used. In particular the results of the models are not applicable to other areas which differ in soil type, topography, and cropping programs.

Management of Water Resources on the Eastern Portion of Long Island

As described in Chapter 4, Long Island is one area in the northeastern United States where the nitrate-nitrogen content of ground water utilized for domestic supplies often exceeds 10 mg NO_3–N/l. The following relates the investigation described in that chapter to some management questions governing the use of the island's water resources.

First, precipitation exceeds evapotranspiration by about 50 cm/yr, so presumably the replenishment of ground water by precipitation is approximately 5×10^6 l/ha/yr. If no more than an average of 50 kg N/ha is dissolved in this water annually, then the concentrations of nitrate-nitrogen in the ground water will be less than

10 mg NO_3–N/l. The fertilization procedure for potatoes, which is recommended in Chapter 4, meets that criterion according to all the evidence now available.

Currently, the recommended annual fertilization rate for grass (lawns, golf courses, etc.) on Long Island is about 175 kg N/ha/yr (4 lbs. of N per 1000 ft^2). Removal of nitrogen in clippings is generally a small fraction of the amount applied. After a period of 10 to 20 years under such management, the soil organic nitrogen content will approach a steady state (inputs of plant residues = decay of organic matter). In the generally acid soils on Long Island, NH_3 volatilization is likely to be small. Thus there is strong circumstantial evidence to support the hypothesis that a substantial fraction of nitrogen applied to grass in lawns, golf courses, etc., ultimately is leached to the ground water as nitrate. Applications of 175 kg N/ha/yr to grass which has been so fertilized for 10 to 20 years may lead to ground water which contains 20 to 30 mg NO_3–N/l. Thus two research needs are clearly evident: (a) an evaluation of the hypothesis that present fertilization schemes for grass are polluting ground water and (b) if this hypothesis is confirmed, devise procedures for fertilizing grass which will not excessively pollute the ground water.

A third source of nitrogen is sewage which is disposed via home disposal fields. Since the nitrogen content of domestic sewage is about 6 kg N/cap/yr, and water consumption is 160 m^3/cap/yr, a population density of 8 persons per ha could supply about 50 kg N/ha/yr to the recharge water and hence produce about 10 mg NO_3–N liter in the ground water, when diluted by the water replenished by precipitation.

The foregoing illustrates that several sources of nitrogen contribute to the nitrate in ground water on Long Island. Control of any one source will not insure ground water of acceptable quality everywhere. This is particularly true with respect to domestic sewage. Removal of nitrogen from domestic sewage and recharge, or zoning to insure less than 8 people per ha, will still leave fertilization of grass as a potential source.

One solution to the water problem on Long Island is to make no attempt to provide water with less than 10 mg NO_3–N/l for all uses, and instead to use bottled water for drinking and culinary purposes. This is an unacceptable alternative to many people; however, consideration of the expense (in terms of energy and money) of insuring that all water which enters homes contains less than 10 mg NO_3–N/l will be expensive and money spent on water supply cannot be used for social programs, health care programs, highway construction, etc. Do the benefits of low NO$_3$–N in *all* domestic water justify the costs?

Another alternative would be to provide strict control of nitrogen inputs to ground water in portions of each town such that the ground water would contain less than 10 mg NO$_3$–N/l. This area would then serve as a source of water for the remainder of the town where much less restrictive controls could be utilized.

FLOWS OF NITROGEN
AND PHOSPHORUS IN STREAMS

Assessment

The measurement of nitrate-nitrogen in the stream appeared to be more straightforward than phosphorus. Concentrations were less variable especially as discharge rates varied. Seasonal changes were identified, but such changes occurred relatively slowly compared to the more abrupt variations observed in concentrations of phosphorus. Given this characteristic of nitrate-nitrogen, it is believed that 20 to 40 samples taken over one year will provide a representative sample at a given point in the stream.

No procedure for identifying sources of nitrate, based on stream composition, was devised. Some of the reasons follow. First, no major point sources could be identified. Second, since nitrate does not react with mineral matter, the water moving through the unconsolidated material above the bedrock will usually contain concentrations of nitrate comparable to that in overland flow.

Based on general agronomic information and variation in nitrate concentrations among subwatersheds, about one-half of the nitrate was estimated to be derived from corn land. The remainder was derived from several sources including domestic sewage.

In the previous section, possible effects of ammonia in the atmosphere were briefly discussed. Studies in these investigations suggest that the impact of atmospheric nitrogen on water quality does not appear to be very large. For example, in the Fall Creek watershed, the output of nitrate from all subwatersheds was less than the estimated inputs of nitrogen in the precipitation (8 kg/N/ha/year) even in those watersheds where 30 to 40% of the land was used for agriculture. In those watersheds least influenced by human activity, losses of nitrate-nitrogen were on the order of 1 kg N/ha/yr, which is in agreement with outputs from a forested watershed reported by Bormann *et al.* (1968). In all subwatersheds, outputs of ammonium-nitrogen in the stream water was much smaller than outputs of nitrate-nitrogen. Thus the nonagricultural segments of the landscape seem capable of removing the inputs of nitrate and ammoniacal forms of nitrogen in precipitation plus contributions from other atmospheric sources.

In the work undertaken on the movement of phosphorus in streams, a great deal of thought was given to the problems of estimation. In considering the flow of phosphorus for management purposes, it is necessary to specify precisely what should be measured, how it should be measured, and when it should be sampled. Some of these problems are presented below.

1. The experience gained in this work suggests that considerable thought and laboratory work need to be carried out prior to any formal collection of stream data. The stream contains several basic chemical forms of phosphorus, not all of which are equally available to aquatic organisms. The complex question of which fractions should be measured and details of the analytical procedures need to be carefully worked out. For the purposes of this study, the basic fractions of phosphorus were determined as shown in Figure 8.1. The principal fraction believed to be immediately available for plant use is molybdate-reactive phosphorus. Loading of the different forms estimated in Fall Creek for an average year are shown in Figure 8.2. As can be seen, the biologically available fraction is probably less than 25% of the total phosphorus in the stream.

2. The phosphorus concentration in stream water varies with discharge rate and with season, the nature of this relation being remarkably variable among subwatersheds. In Fall Creek, about half of the phosphorus transported annually from the watershed is removed in about 10% of the time during which highest discharge rates are occurring. This is likely to be typical of many other streams. An estimate of annual loading, even with a standard error equivalent to 25% of the loading, requires continuous flow records at a gaging station whose characteristics are stable and adequately calibrated, and at least 5 to 7 samples should be taken during each high discharge rate period. The sampling of these events should be well distributed over the season during which high discharge rate events are most likely, in addition to 30 to 60 samples taken during low- and medium-discharge rate intervals.

Regular or systematic sampling is unlikely to provide a reliable measure of the total loads carried during the high discharge rate intervals unless the frequency of sampling is sufficiently high to obtain a representative sample of the high flow events. With such

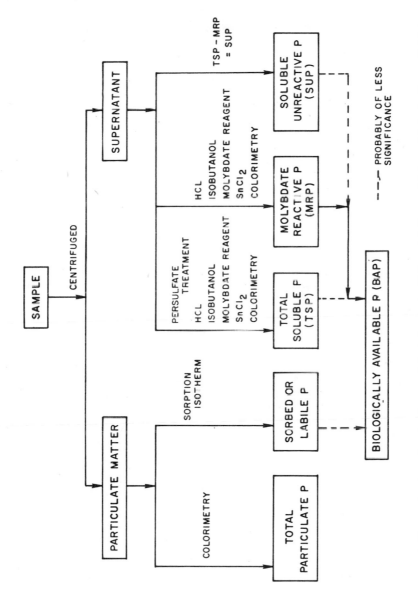

Figure 8.1. Determination of major forms of phosphorus to distinguish those which are biologically available.

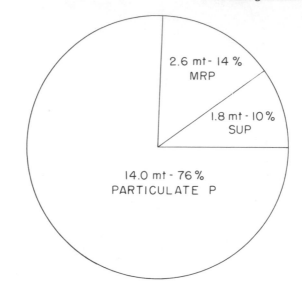

Figure 8.2. The estimated amounts of various fractions of phosphorus carried out of the Fall Creek watershed each year (post-detergent ban). The molybdate reaction P is primarily in the dissolved inorganic form while the soluble unreaction P is primarily in the dissolved organic form.

frequent, regular sampling, the program will incur redundancy be-cause many more samples at low discharge rates will be taken than are necessary to characterize the low flow.

3. Since a major aim of these studies was to attribute observed levels of phosphorus in the stream to sources created by human activity in the watershed, it was necessary to estimate the amount of phosphorus expected in the absence of human activity. This can only be done by indirect methods such as those described in Chap-ter 3. The procedure used was to measure biologically available phosphorus in several small subwatersheds currently devoid of appreciable human activity and in wells reasonably distant from sources such as barnyards that could affect the quality of water in the wells. Because the wells and streams are fed by water moving through the unconsolidated material above the bed rock, these samples were regarded as being good indicators of background levels of phosphorus. This unconsolidated mantle is in fact a gi-gantic sink for phosphorus and only very large phosphorus inputs are likely to appreciably influence the bulk of the water moving through it. There are two exceptions: (a) Where water moves down-slope in the plow layer, because of underlying less permeable soil

material, the concentration may be fairly high if the plow layer has received large amounts of manure or fertilizer phosphorus. This may be the explanation for high concentrations of phosphorus sometimes found in tile drains. (b) Drained organic soils have relatively small amounts of mineral matter available for reaction with phosphorus, and hence water leaving these soils may have very high phosphorus concentrations derived from fertilizer and decomposition of organic matter.

4. Determination of the original sources of phosphorus after it has entered the stream is practically impossible, except in a very crude way. The hydrograph analysis procedure described in Chapter 3 has the disadvantage that the results obtained are highly ambiguous. Concentrations during high flows are influenced by point source phosphorus stored in the stream bed during low discharge as well as by phosphorus from diffuse sources transported to the stream by overland flow. Careful selection of subwatersheds in which there is primarily farmland and other potential diffuse sources is essential to estimation of such sources. Unfortunately, there are likely to be point sources in most subwatersheds even though carefully selected. However, these sources can be located by sampling at frequent intervals along streams during periods of low flow.

The phosphorus from point sources was transferred from the water to the bottom sediment as it moved downstream in Fall Creek during low discharge rate intervals. Some of this phosphorus was retained in a labile form on the sediment and redissolved during high discharge as the bottom sediment was swept up in the turbulent water, and the remainder was converted to a relatively inert form which presumably has a low biological activity. This conversion of soluble phosphorus from point sources such as sewage to relatively insoluble forms is an important mechanism from the viewpoint of nutrient management. It suggests that the enrichment of lakes could be ameliorated if point discharges were located in the drainage system as far from the lake as possible. However, validation of such a conclusion for other watersheds is desirable.

The following considerations are important in extrapolation from one watershed to others. The data presented in Chapter 3 illustrate that a fairly small number of human activities have a major influence on phosphorus carried in streams (e.g., sewage treatment plants, area and location of barnlots, area of poorly drained soils receiving manure). Thus a relatively small change in the proportions of these activities among different watersheds may produce major errors in extrapolation. Furthermore, many of the important parameters are not amenable to estimation with available data, or data which are easily acquired. Thus models which take these variables into account are of limited usefulness since the data are not available. The usual compromise leads to selection of independent variables which are readily estimated although not necessarily the most significant.

Management

In the case of managing nitrogen and phosphorus in streams, it is perhaps arbitrary to distinguish among assessment, scientific and management objectives since all are highly interwoven. The preceding discussion implies that a primary aim of management is to delineate the objectives of a sampling program. These may include measurements which are intended to:

1. Monitor the flow and quality of the river water for the purpose of maintaining standards;
2. Obtain information for purely scientific investigations
   a) to learn about the processes within the river,
   b) to measure substances within the river so that they can be attributed to sources on land;
3. Obtain information upon which predictions can be based for the purpose of determining future management policies.

These objectives have specific requirements regarding the number and frequency of sampling. For example, a scientific study to assess the impact of some event may require intensive sampling of the river over a short period. For general monitoring purposes, less frequent but periodic samples may be more appropriate. It is difficult to specify general requirements because they will be determined by the basic characteristics of the river system being investigated.

Concerning objective 2b, a great deal of detailed information was obtained about the Fall Creek watershed in the investigations described in the previous chapters. As outlined in Chapter 3, the data were used to attribute fractions of the observed levels of phosphorus in the creek to their original sources on land. Except for some well-defined point sources of phosphorus such as one barnyard or the Dryden sewage treatment plant, the uncertainties associated with the estimates were large.

A conclusion that may be drawn from this is that the quantification of non-point sources of phosphorus, or other substances, is nearly impossible on the basis of stream samples only. Rather than such indirect attempts to identify and quantify sources, it is necessary to make direct observations of the sources as they occur in the landscape and of the transport of material to the stream in conjunction with the observations made in the stream. Thus there are four essential ingredients in assessment for management purposes: (a) observations of stream composition, (b) observations of activities in the watershed, (c) appropriate field measurements of representative sources, and (d) logical integration of the above.

This conclusion has significance for national policy which requires the identification and control of non-point sources of pollution in surface water. For example, in section 208 of PL 92-500, it is stated that area-wide waste treatment management plans should include a process "to (1) identify, if appropriate, agriculturally- and silviculturally-related non-point sources of pollution, including runoff from manure disposal areas, and from land used for live- stock and crop production, and (2) set forth procedures and methods (including land use requirements) to control to the extent feasible such sources." The second part of this statement implies management and assumes that the first part, assessment, can be accomplished. From the work discussed in this book, it is evident that assessment is a time-consuming and expensive process. The usefulness of the control measures studied in Chapter 5 in terms of reducing phosphorus from agriculture which is carried in Fall Creek is highly dependent on quantitative knowledge of non-point sources.

Implementation of legislation to monitor and regulate non-point sources of pollution is dependent on data which is not currently available for most watersheds. It may be sufficient to estimate losses of substances based on a sample set of land characteristics and use, and to associate these with observations in surface waters. It may then be possible to relate observations in other streams to similar land conditions without necessitating direct measurement of the latter in all watersheds. If generalization from a small sample is not possible, and it is not yet known whether the contrary is true, assessment and management of non-point sources of pollution on a national scale will be difficult to accomplish.

In summary, the foregoing discussion illustrates the difficulty and expense of a meaningful program to assess and monitor diffuse sources and numerous small point sources widely dispersed such as are found in most predominantly rural watersheds.

NITROGEN AND PHOSPHORUS IN LAKES

Assessment

Evidence suggests that rivers draining agricultural areas receive large inputs of nitrate-nitrogen from such areas (Harmeson and Larson, 1969; Porter and Glassey, 1975). Should the river subsequently discharge into lakes, it follows that a significant quantity of nitrate-nitrogen within the lakes should be attributable to surrounding agricultural land.

However, all available evidence supports the hypothesis that nitrogen is not a limiting substance in northeastern lakes and hence has little significance for management. This also applies to lakes that are used as a source of potable water because no lake investigated had levels of nitrate which approached 10 mg NO_3–N/l.

Concerning the assessment of phosphorus in lakes, the discussion in Chapter 2 indicated that many limnologists measure total phosphorus in lake water. The particulate matter and associated phosphorus is deposited on the lake bed under the relatively quiescent conditions in lakes. Therefore, total phosphorus in lakes is virtually equivalent to biologically available phosphorus in streams.

Considered simplistically, biologically available phosphorus entering a lake remains in the water column for several months. During the summer, much if not all of this phosphorus becomes incorporated into the phytoplankton. Relations between phosphorus and phytoplankton for a group of New York lakes were developed in Chapter 2, and it is postulated that these or similar relations are generally applicable to lakes throughout the north temperate regions of the world.

According to the evidence in Chapters 2 and 3, biologically available phosphorus (BAP) in streams provides more meaningful loading data than does total phosphorus. The use of BAP rather than total phosphorus alters the relative significance of various sources. The ratio of BAP to total phosphorus is not constant. This ratio is not only variable for land runoff, but ratios for land runoffs are usually lower than the corresponding ratios for domestic sewage. For example, in Fall Creek about 75% of total phosphorus was in particulate form, much of which was derived from farmed land according to calculations based on the universal soil loss equation. Less than 25% of the BAP was attributed to agriculture. If loading had been based on total phosphorus, farmed land would have contributed much more than 25%. On the other hand, domestic sewage was the source of only 6% of the total phosphorus but about 25% of the BAP.

Concerning the sources of phosphorus in lakes, it is realized that every landscape is unique, and at present there is no way of estimating the variability in phosphorus inputs to lakes that differences in slope, soil type, vegetative cover and climate produce. Nevertheless, the coefficients representing various sources of phosphorus in the lakes in central New York may be extrapolated with caution to other watersheds in the glaciated areas of the northeastern United States, southeastern Canada, and perhaps some of the upper midwestern United States. Natural vegetative cover and agricultural practices are similar over these regions.

The watershed characteristics which were selected in Chapters 2 and 3 as independent variables for estimating phosphorus delivery to lakes were (1) sewered population, (2) unsewered population, (3) agricultural land, and (4) nonagricultural land (all expressed on a per unit of lake surface area). The data necessary for compilation of these parameters for each lake basin were readily available, while data for other parameters such as barnlots were not. All four sets of variables were correlated with levels of summer chlorophyll and winter phosphorus in the lakes, and with each other. Calculations of phosphorus loading, by assigning an estimated input per unit per year to each characteristic and then summing over the whole lake basin, was also well correlated with the lake parameters. Superficially, this correlation indicates that the values assigned to each basin parameter were reasonably accurate and hence useful for management purposes. However, the issue is clouded when the important lake basin characteristics are correlated with each other. This problem is discussed further in Chapter 3.

It should be noted that the factors which control macrophytes may be entirely different from those which control algae. Rooted macrophytes may derive most of their essential nutrients from the bottom sediments. Other macrophytes may thrive in shallow water where the exchange of nutrients between water and sediments may be reasonably effective.

Management

Management of a lake will be primarily dictated by its use. Two principal uses are:

1. recreation including fishing, boating and swimming.
2. withdrawal of water for domestic, industrial or agricultural use.

A third use which is incompatible with present laws is to use the lake as a recipient for untreated wastes.

It is now commonly accepted that recreational opportunities are a resource, and their development, allocation and use should be economically evaluated with other uses, and resources. The aim of much recent cost-benefit research in particular has been to establish means of assuring an optimal development of resources with multiple uses. Unfortunately, no cost-benefit methods are entirely satisfactory when applied to recreation. A major part of the difficulty is that many recreational values are not determined by market activity. Much of the outdoor recreation in the United States has been characterized by public ownership and low-cost access by the public. In many such cases, nothing resembling a true market mechanism exists wherein the value of the recreational experience might be reflected in a market price. (Stoevener et al., 1972)

As a result of these difficulties, attempts have been made to measure the value of recreational areas by indirect means. In a review, Crutchfield (1962) distinguishes two groups of methods, one being unsound and deserving complete rejection, the other being analytically correct but very difficult to measure. Among the former, Crutchfield includes methods such as the costs of producing the resource, and the opportunity cost of the time spent in recreation. The correct methods involve the derivation of a demand schedule for the recreation which provides an estimate of the net economic yield of the recreational area. The schedule is usually calculated by estimating the differential costs incurred by usage of the recreational site and by calculating how these affect the number of visits made (Knetsch, 1963; Smith, 1972). Unfortunately, the computation of differential costs is usually arbitrary. For example, a value is usually attributed to the amount of time taken in travelling to the site. An immediate difficulty is that for many people, the journey itself may incur pleasure which should not be attributed to the recreational experience at the final destination. Also the value of time doubtless varies between individuals, particularly as the required time to travel itself varies.

A further problem is that different recreational uses themselves may be in conflict. For example, swimming and water skiing may not be compatible in the same lake. Further, as argued in Chapter 2, a reduction in the enrichment of a lake may increase the transparency of the water and hence its aesthetic appeal. Unfortunately, the lower nutrient levels will also probably reduce the number of higher organisms such as fish supported by the enrichment, hence with disappointment for the fishermen. The latter may not be consoled with his smaller catch, by the reflection that the greater transparency allows him to at least see that the fish have gone.

A major part of the benefits of changing the levels of phosphorus in lakes would stem from their recreational uses. Such benefits can only be derived by estimation of demand schedules. These investigations were unable to consider recreational benefits in detail, but there is clearly great need for such research.

Many studies have been reported concerning the effect of lake enrichment upon the use of the lake for water supply (Baylis, 1955; Gamet and Rademacher, 1960). The main problem for the water engineer is that filtration in a water treatment plant becomes inefficient and expensive if the water has a high concentration of algae. As a result of the algae, short filter runs occur which are expensive because they reduce the effective capacity of the plant. In general, relations between lake enrichment and the cost of water treatment have not been fully established. However, with the increasing use of limited water resources, such costs may come to be an important factor in management.

Finally, lake enrichment due to the activities of man in lake watersheds is often described as being a process which accelerates the natural aging of lakes. Perhaps from this the impression has been fostered that the consequences of lake enrichment are irreversible—for example, the alarm expressed in recent years by the mass media about Lake Erie. However, considerable evidence shows that if nutrient loadings to lakes are reduced, they can in large measure be restored to their former condition (Chapter 2).

MANAGEMENT APPLICATION TO
LAKES IN CENTRAL NEW YORK STATE

In this section, some of the results of previous chapters will be applied to three lakes, Canadarago, Cayuga and Skaneateles, in central New York. The estimated phosphorus loading under five management policies was calculated for each lake. This loading was then substituted into the regression of chlorophyll on biologically available phosphorus loading, calculated from the measurements reported in Chapter 2, to provide an estimate of the chlorophyll levels which would be expected to follow each policy. The five policies assumed were:

I. No ban on phosphorus in detergents;
II. Ban on phosphorus in detergents;
III. In addition to Policy II, 80% of the phosphorus from unsewered wastewater removed by treatment;
IV. In addition to Policy III, 50% of the phosphorus from unsew-

ered wastes and 50% of the phosphorus originating from agriculture removed;

V. All human sources of phosphorus removed by elimination of human activity in watershed.

Expected consequences of each policy are illustrated in Figures 8.3 and 8.4.

The sequence of policies assumes that following the first policy with no explicit control, the least expensive source is controlled first, followed successively by the next least expensive source. It was argued previously that after banning phosphate detergents, the least expensive source of phosphorus to control was that in sewage. Its removal in tertiary treatment was estimated to cost from $5 to $12 per kg of phosphorus per year depending on the size of the plant. An average cost may be assumed to be about $8/kg P for central New York, which is equivalent to $4/cap/yr.

Removal of phosphorus from the wastewater of the currently unsewered population by tertiary treatment requires a system of collection and secondary treatment. Not all of the cost of such a system should be charged to phosphorus removal because of other benefits such as reduced health hazards from pathogens. It was estimated that the cost of removing half of the BAP reaching the lakes from unsewered populations would be higher than for sewage from existing plants. This is because only a fraction of the phosphorus in the sewage from these systems is presently entering lakes as BAP. Collection and tertiary treatment to remove phosphorus will not reduce inputs as much, on a per capita basis, as removal of phosphorus from the sewage of the presently sewered population. Costs were estimated at $15 to $50 per kg of P.

If the BAP losses from agriculture are primarily from barnlot runoff and manure applied to the more poorly drained soils, then these sources should be the least expensive agricultural sources to control. Barnlot runoff control plus a requirement that on the more poorly drained soils, manure either be incorporated into the soil immediately after spreading or spread only on meadow crops during the summer could cost farmers $30 to $150 per kg of BAP prevented from entering lakes.

No attempt was made to estimate the cost of reducing the phosphorus inputs from human activities to zero. So far as known, there is no practical means of doing this. Removal of all human activity, as assumed in policy V, seems the only way to achieve such a reduction, although this is both drastic and impractical.

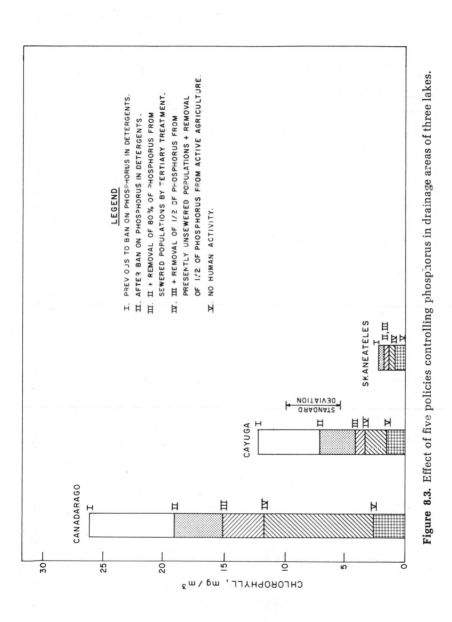

Figure 8.3. Effect of five policies controlling phosphorus in drainage areas of three lakes.

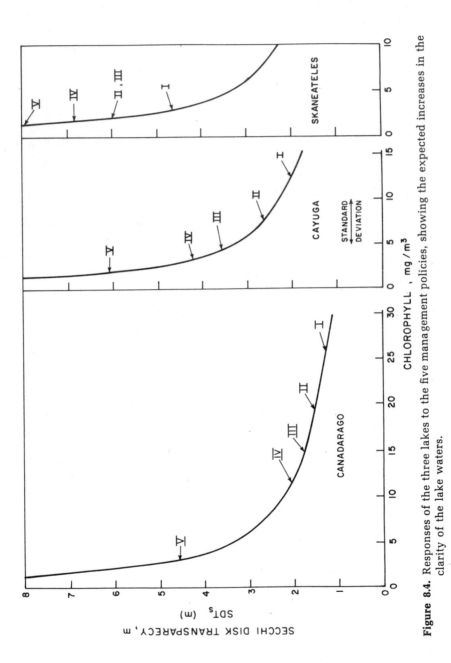

Figure 8.4. Responses of the three lakes to the five management policies, showing the expected increases in the clarity of the lake waters.

Policy I. Policy I represents the conditions existing prior to the time when the ban on phosphorus in laundry detergents imposed in June, 1973 would have had an appreciable effect. The relations between chlorophyll in lakes and the inputs of BAP from the lake basins were developed from these measurements, utilizing the results with all 13 lakes in the central New York region.

The BAP from the sewered populations prior to the ban on phosphorus in laundry detergents was assumed to be 1.2 kg BAP/cap/yr. This is based on an estimated content of 1.5 kg P/cap/yr in the original sewage with 20% removal during treatment and no loss between the sewage treatment plant and the lake. Second, the BAP from the unsewered population was estimated to be 20% of 1.5 kg P/cap/yr. Third, the BAP from active agriculture was estimated at 18 mg/m³ of annual stream flow from the fraction of the lake basin in active agriculture. The biogeochemical BAP was estimated at 15 mg/m³ of stream flow. The average stream flow was taken from U.S. Geological Survey records. Using these parameters and the lake basin characteristics, the inputs of BAP (kg/ha of lake surface) were calculated from equation 8.1.

$$BAP = 1.2\,P_s + 0.3\,P_u + 15\,DA_T + 18\,DA_A \qquad (8.1)$$

where,

BAP is in kg/ha of lake surface/yr
P_s sewered population, per ha of lake surface
P_u unsewered population, per ha of lake surface
D is average annual stream flow, $m^3/ha \times 10^{-6}$
A_A is active agricultural land, ha/ha of lake surface
A_T is total land area, ha/ha of lake surface.

These values of BAP were then used as independent variables and the observed values of chlorophyll in the 13 lakes reported in Chapter 2 were used as dependent variables to calculate the following regression equation:

$$Chl = 0.2 + 27.9\frac{BAP}{Z} \qquad r^2 = 0.71 \qquad (8.2)$$

where,

Chl is summer chlorophyll, mg/m^3
BAP is defined by equation 8.1, which differs from the similar equation in Chapter 2; and
Z is the effective depth of mixing (see addendum to Chapter 2).

Equation 8.2 was used in all five policies to estimate chlorophyll if a given loading were continued for a period of several years. The chlorophyll calculated from 8.2 for each of the three lakes chosen as examples is illustrated by point I on the bars in Figure 8.3 and curves in Figure 8.4. Figure 8.4 utilizes the relationship between transparency and chlorophyll which was developed in Chapter 2.

Policy II. Next, the influence of the ban on phosphorus in laundry detergents was estimated by reducing the inputs of P_s and P_u by half those listed for Policy I. Results reported by Hopson (1975) indicate that a 50 to 60% reduction in phosphorus content in domestic sewage influence was obtained in three large plants in the Buffalo area by a similar ban in Erie County, New York which was instituted in 1972. Thus the coefficient in equation 8.1 for P_s became 0.6 and for P_u became 0.15 with the other two coefficients remaining the same. These values of BAP were then substituted into equation 8.2 to calculate the chlorophyll levels expected if phosphorus inputs into the lakes remain at these levels for a long period of time. These points are labeled II in Figures 8.3 and 8.4.

Policy III. Point III in the diagrams was based on estimated inputs if the existing sewage treatment plants were modified to include phosphorus removal to the extent that only 0.12 kg P/cap/yr remains in the effluent. The other coefficients remained the same as those for Policy II.

Policy IV. The next restriction was to construct sewers for the portion of the unsewered population which is both (a) close to the lakes, and (b) where a large fraction of the individual disposal fields do not function properly. In the absence of definite data, the net effect is assumed to reduce by 50% the inputs to lakes of phosphorus from the presently unsewered population. In addition, the inputs from active agriculture were reduced by 50%. The resulting loading values were substituted into equation 8.2 and the calculated chlorophyll levels were plotted as point IV in Figures 8.3 and 8.4.

Policy V. Finally, the levels of chlorophyll were calculated as above by reducing all inputs from human activity to zero so that only biogeochemical inputs remained. These are plotted as point V in Figures 8.3 and 8.4.

While the reader may be skeptical about the result of the above extrapolations, the following comments are offered in justification for the calculations. First, in our judgment, this represents the most reasonable interpretation of what is known about the relationships between phosphorus inputs from human activities and chlorophyll in the lakes in central New York. Second, any major improvements in reliability of the relations will only result from the collection of more data. Refinements of the methods of analysis of the data now available might perhaps improve the relationships somewhat, but such refinements are not likely to make a major change in the general conclusions.

From the above it is apparent that (a) an undesirable degree of uncertainty is associated with the relations between human activities and chlorophyll levels in lakes, and (b) this uncertainty is unlikely to change unless considerably more research, which requires time and money, is carried out. The question is, therefore, what is a reasonable course of action?

First, considerable evidence is now available which indicates that over a period of years, the phosphorus available to algae is not conserved; that is, high levels of algal growth appear to depend upon continuing inputs of biologically available phosphorus. The classic example of Lake Washington is illustrated in Figure 2.6, page 40. Effects of a particular source of phosphorus appear to be reversible whenever a control measure is implemented. Therefore, delay in introducing controls will probably not cause irreversible changes in the lakes.

The reversibility of changes suggests an evolutionary policy in which controls are gradually instituted on a source by source basis. The relations developed in Chapters 2 and 3 suggest in addition that information obtained about one lake could be applied to others.

In effect, a first step of the evolutionary process is now in progress in that the ban on phosphorus in detergents was instituted in 1973. A second step is underway in Canadarago Lake where in 1973 the Richfield Springs primary sewage treatment plant was replaced by a plant which includes facilities to remove phosphorus. Similar facilities are being installed elsewhere in the central New York region. As the detergent ban and the removal of phosphorus from sewage take effect, it is essential that corresponding changes in the lakes be adequately monitored and assessed. The knowledge gained thereby could be applied in the formulation of further management policy, not only for the lakes directly affected, but similar lakes elsewhere.

If policies were based on such systematic use and transfer of information between lakes, a framework would be established whereby the costs of reducing nutrient inputs to lakes would be assessed. For example, Figure 8.5 shows the relation between Secchi disk transparency and chlorophyll in the lakes of central New

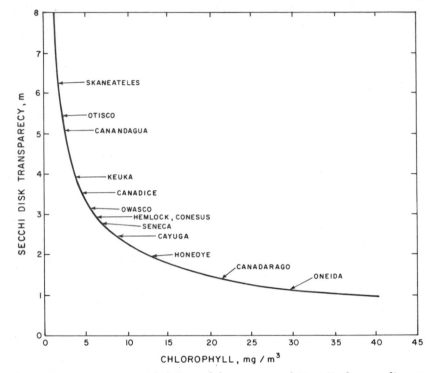

Figure 8.5. Present state of thirteen lakes in central New York according to the Secchi disc transparency—chlorophyll relation.

York in the period prior to the ban on phosphatic detergents. Using this relation, the costs could be stated in the terms that it would probably cost X dollars to transform conditions in lake Y to those existing in lake Z. This would provide a graphical basis for evaluating the effects of management policies proposed for the lakes. Given appropriate data, further relations between the levels of phosphorus input to lakes and their use, such as for fish production and water supply, could be derived and used in an analogous manner.

CONCLUSIONS

These studies had two main aims: 1) to consider the effects of man's activities on the flows of nitrogen and phosphorus, especially in rural areas, and 2) to consider management of the two nutrients with respect to food production and the environment so that welfare of society is increased.

Nitrogen and phosphorus contained in fertilizer and animal and human wastes are actual or potential resources. Their management also involves the management of other resources such as soil as a factor in food production and ground and surface water as possible recipients and carriers of the nutrients.

Either nitrogen or phosphorus may have adverse environmental effects on water or the atmosphere, and in a wide sense the conservation of resources and preservation of the environment are complementary problems. It was an implicit assumption in this work that the management of the nutrients be considered in the broad framework of resource management rather than in the narrower terms of pollution control.

This study attempted to quantify the flows of nitrogen and phosphorus across the landscape, especially as they are affected by man's activities, to assess some of the consequences of these flows in final receiving waters. The core of the investigations was a highly detailed study of a specific watershed, Fall Creek, and a study of 13 lakes in central New York. Other investigations included a study of nitrate in ground water on Long Island and a study of the control of nitrogen in animal manure. Although detailed knowledge obtained from these investigations was specific to New York State, many of the results have general as well as local applicability. Since the movement and effect of the nutrients in the environment were quantified, alternative management policies together with associated costs and benefits can now be better evaluated.

It was possible to identify and to estimate the nature of each important source in the flow and balance of both nitrogen and phosphorus in the Fall Creek watershed. With respect to the level of nitrate-nitrogen in the creek, concentrations rarely were above 3 mg/l, well below the drinking standard of 10 mg/l. This was the case with other watersheds and lakes in central New York, and given that nitrogen is not a limiting nutrient for algae, it would appear that there would be little benefit in managing uses of nitrogen on land to reduce inputs to the above surface waters.

On the other hand, in parts of Long Island, the concentration of nitrate-nitrogen is increasing and in some aquifers exceeds 10 mg./l. Since ground water aquifers are the only present source of water supply for the eastern part of the island, alternative management policies should be developed. Some possible alternatives, which would reduce the concentration below 10 mg/l, include lowered applications of nitrogen to crops and lawns and a population density restricted to less than 8 persons/ha in areas from which water will be drawn.

Principles developed and discussed in this work are applicable to other areas where surface and ground water contain unacceptably high levels of nitrogen. As discussed above, alteration of timing and rates of nitrogen fertilization of crops, control of manure application rates and of nitrogen in animal manures before land application, control of nitrogen in domestic sewage effluent, and limits on population density may be appropriate management strategies.

The effects of phosphorus on algal productivity were quantified. It was determined that it is important to consider that fraction of phosphorus which is biologically available. Ways by which levels of phosphorus could be reduced in surface waters were also studied. It was estimated that, following the ban on phosphates in laundry detergents, methods of controlling inputs of biologically available phosphorus to lakes, in ascending order of costs, were tertiary treatment of domestic sewage, control of barnyard runoff, control of land runoff by altering manure practices and by changing from row crops to non-row crops. However, the costs of controlling the agricultural sources of phosphorus were very much higher than the costs of controlling phosphorus in domestic sewage, and control of the agricultural sources is probably less effective.

In brief, the following specific results were obtained:

1. Relations between algal productivity and biologically available phosphorus transported into lakes were derived.

2. Levels of sediment, phosphorus and nitrogen in a stream were quantified and related to human activities in the watershed.

3. Consequences following application of nitrogen and phosphorus to soil were evaluated, especially with respect to ground water.

4. Economic effects of phosphorus controls in farming were determined.

5. Methods were developed to allow the control of nitrogen in animal manures prior to their disposal.

6. Social attitudes to problems of water quality, and ways in which the public can respond to such problems were assessed.

7. Some guidelines and uses of the results for management of nitrogen and phosphorus were proposed.

The above in total provide a comprehensive means of assessing and specifying management alternatives in controlling the effects of using nitrogen and phosphorus. However, the implementation of such alternatives will depend on the broader issues of allocation of available means and resources. For example the preservation of a water resource whose use is jeopardized by high levels of nitrogen or phosphorus has to be compared with other social objectives of high priority. In the end, whether the uses of the nutrients are more stringently controlled becomes an issue of deciding which management policies with respect to resources most promotes community, state and national welfare.

Considering the complexity of the research reported above and of the environmental issues related to nitrogen and phosphorus in food production, the above conclusions do not completely summarize the research or its implications. The interdisciplinary team prefers to have the investigations stand on their own merit and to let the reader utilize them to draw further conclusions.

REFERENCES

Ashraf, M. and R. L. Christensen. 1974. An analysis of the impact of manure disposal regulations on dairy farms. American Journal of Agricultural Economics 56:331–336.

Baylis, J. R. 1955. Effect of microorganisms on length of filter runs. Wat. Works Eng. 108:127.

Bormann, F. H., G. E. Likens, D. W. Fisher, and R. S. Pierce. 1968. Nutrient loss accelerated by clear-cutting of a forest ecosystem. Science 159:882–884.

Buxton, B. M. and S. J. Ziegler. 1974. Economic impact of controlling surface runoff from U.S. dairy farms. Agricultural Economic Report No. 260. Economic Research Service, USDA, Washington, D.C.

Casler, G. L. and E. L. La Due. 1972. Environmental, economic and physical considerations in liquid handling of dairy cattle manure. New York's Food and Life Science Bulletin No. 20. Cornell University, Ithaca, N.Y.

Crutchfield, J. A. 1962. Valuation of fishery resources. Land Economics 38:145–154.

Gamet, M. B. and J. M. Rademacher. 1960. Study of short filter runs into Lake Michigan water. J. Amer. Wat. Works Assoc. 52:137–152.

Good, D., L. J. Conner, J. B. Johnson, and C. R. Hoglund. 1973. Impacts of imposing selected pollution controls on Michigan dairy farms. Michigan Farm Economics, Michigan State University, East Lansing, Michigan.

Harmeson, R. H. and T. E. Larson. 1969. Quality of surface water in Illinois, 1956–1966. Illinois State Water Survey, Urbana. Bulletin 54.

Hidy, G. M. 1973. Removal processes of gaseous and particulate pollutants. In Chemistry of the Lower Atmosphere, ed. by S. E. Rasool. Plenum Press, New York, pp. 121–176.

Hopson, N. E. 1975. Phosphorus removal by legislation. Water Resources Bulletin. American Water Resources Association 11:357–364.

Klausner, S. D., P. J. Zwerman, and D. R. Coote. 1975. Design parameters for application of dairy manure. Terminal Report on Project No. 800767. National Environmental Research Center, EPA, Corvallis, Oregon.

Knetsch, J. L. 1963. Outdoor recreation demands and benefits. Land Economics 39:387–396.

2

2

ᴀ ok stop

Likens, G. E. 1972. Chemistry of precipitation in the Finger Lakes region. Tech. Report No. 50. Cornell University Water Resources and Marine Science Center, Ithaca, N.Y.

Likens, G. E. 1974. Runoff of water and nutrients from watersheds tributary to Cayuga Lake, New York. Tech. Report No. 81. Cornell University Water Resources and Marine Science Center, Ithaca, N.Y.

Lund, L. J., D. C. Adrfano, and P. F. Pratt. 1974. Nitrate concentration in deep soil cores as related to soil profile characteristics. J. Environ. Quality 3:78–82.

McConnell, J. C. 1973. Atmospheric ammonia. J. Geophysical Res. 78:7812–7821.

Nassau-Suffolk Regional Planning Board. 1968. Existing land use. Hauppage, N.Y.

Porter, K. B. and C. R. Glassey. 1975. Statistical estimation using time-related data with applications to the study of inorganic nitrogen in the system of the River Lee. Water Resource Center, Stevenage, England.

Smith, R. 1972. Recreation benefits: angling. The Trent Research Programme, Vol. 10. Water Resources Board, Reading, England.

Stoevener, H. H., J. B. Stevens, H. F. Horton, A. Sokoloski, L. P. Parish, and E. N. Castle. 1972. Multi-disciplinary study of water quality relationships: a case study of Yaquina Bay, Oregon. Oregon Agricultural Experiment Station, Corvallis, Oregon.

Taylor, C. R. and E. R. Swanson. 1975. The economic impact of selected nitrogen restrictions on agriculture in Illinois and twenty other regions of the United States. Agricultural Economics Research Report 133, University of Illinois, Urbana, Illinois.

Tseng, T. 1975. Ph.D. thesis in preparation. Dept. of Agricultural Engineering. Cornell University, Ithaca.

POSTSCRIPT

Interdisciplinary Research
In the University

INTERDISCIPLINARY RESEARCH
IN THE UNIVERSITY

Among the various forms of interdisciplinary research organization several types are notable in the university environment. For example, there is the loosely-structured organization such as an institute often involving more than one university where researchers under a grant meet annually to report and correlate their activities. Another type is a collectivity of individuals usually on one campus, linked together by research funds, working under a broad mandate that allows each to pursue his/her own interest with final results published under one cover. Here again such a group may not have been systematically drawn together except for the final report procedures. In many land-grant universities, certain commodity departments may be staffed with faculty representing various related disciplines. Here it is not unusual for a senior faculty member to develop a large research project and then contract with individuals representing various disciplines to carry out selected objectives of the project. Still another type and the one used for this project is a tightly-managed group working under a project director, holding regular and frequent meetings, effecting divisions of labor evolving project designs, decisions and integrating results into a coordinated product.

The funding for this project was provided by the Rockefeller Foundation on a three-year grant with the flexibility to spread the expenditure over a four-year period. From a time perspective the length of the funding periods was fortuitous. Had it been shorter the various disciplines would not have been able to maximize their individual results, nor more importantly blend those contributions into a holistic treatment of the problem. For example, the research team was able to interpret the final report to the extent that agreement was reached on each major point of issue before the final draft was released.

Someone has likened real interdisciplinary research as being closely akin to making a batch of pickles. There is no way that a little vinegar can be hastily squirted on a cucumber, rather it requires a lot of soak time to obtain the desired results.

The interdisciplinary team was composed of graduate students, research associates and professors. The team varied in size in its meetings from six to eighteen. They met a minimum of twice a month during the first two and a half years and weekly during the last year and a half up to publication time, which called for blocks of concentrated time to be set aside.

The team was composed of representatives from six disciplines as departments, namely, Agricultural Economics, Agricultural Engineering, Agronomy, Natural Resources, Poultry Science and Rural Sociology. Members of the team were housed separately in their individual units; communication was achieved through meetings, written materials, telephone interchanges and published resources. A team leader and headquarters were designated from the outset which served as a major basis for planning and implementing team progress and achieving project success.

In retrospect, one of the research associates described his initial experience of becoming involved as a member of an interdisciplinary research team as something like being thrown into a group of cantankerous alligators. The early experience appears to yield considerable manifestation of snapping and contesting for territory in terms of one's own place in the group and the place of one's discipline, in relation to the principles and concepts of other disciplines. There are many processes at work which when analyzed, lend insight into the features of turbulence in the early stages of interdisciplinary work.

Two basic processes are involved, each of which is complex in itself, but which are exacerbated when one is imposed on the other. The first derives from the area of "group dynamics" and proceeds from the well-documented exploratory behavior anticipated in new groups and the subsequent processes that occur from the first tentative, cautious interactions through to the stages of trust, confidence, kidding relationships and cohesion (Hare et al., 1955). The second is the tough set of problems and processes an interdisciplinary project poses, which are not wholly resolved until the final steps of the scientific method are accomplished. The purpose of this postscript is to illustrate the ofttimes independent and sometimes overlapping crucial features of these two processes. In addition, some of the unique benefits that accrue to participants, particularly in the area of training along with perceived advantages and disadvantages, are discussed.

LEADERSHIP: FORMATION AND DEVELOPMENT OF AN INTERDISCIPLINARY RESEARCH TEAM

The leader constitutes a major input source in the formation and development of any new group. This fact is particularly true in the case of a new interdisciplinary group. New groups are helped when they are formed around a common interest. The commonality of interest in a new interdisciplinary group generally emerges later after many other matters are clarified. The motivations for members to participate in an interdisciplinary project represent interests in association with subject matter content, funding resource, dissertation topic, training opportunity or a significant research problem. Ideally the project leader should not have a vested interest in the funds for carrying out his own phase of research in the project. He should be unencumbered to equitably administer the funds required by each disciplinary subgroup based upon predetermined budgets and programs of work.

A feature that makes the project leader's role difficult is that a new group requires many of its needs to be met almost simultaneously. An urgent one for example is the according of necessary recognition and identity to each participant. The gaining of such recognition is a function of at least two things. First, accommodation to the differential statuses that participants, i.e., full professors, research associates and graduate students, bring to the group setting. This is sometimes referred to as the pecking order of the group, the working through of who gives deference to whom. The second is the matter of the language problem which evokes the use of professional jargon characteristic of each discipline as the participants struggle with the first phases of clarifying the dimensions of the research problem.

A common behavioral feature of participants in new groups is the exhibition of uneasiness and anxiety as members jockey for recognition and status while seeking to define the boundaries and parameters of the project. The evidences of stress are raised voices, disparaging remarks about the value of particular contributions, inability to appreciate the relevance of other discipline approaches or impatience with the time it takes one discipline to generate data needed by another. A real task for the leader is to recognize the processes at work and to strive to maintain a forum of understanding which allows for the identification and establishment of each individual's place in the group by means of recognizing and facilitating the legitimacy of each discipline's piece of the research turf.

An interesting phenomenon about the process of gaining individual and discipline recognition is that the principal medium is through the research language spoken. In the physical and biological sciences the terminology tends to derive more from a common ancestry with more commonality of meanings than is the case with the social sciences. An example of this diversity, which posed an observable communications problem for the interdisciplinary research group, was the term "unit of analysis." To the physical and biological scientists this term had a connotation of measurement expressed in numbers or equations. To the social scientists the reference meant more the nature of the object being measured such as the individual, the farms, the community or the watershed. The challenge for a leader and the members is to be aware that while the language is the facilitator it can also create static in the communication system. In the early stages the interdisciplinary team needs to exercise patience to understand and resolve the peculiarities of language, terminology, definitions and meanings. This process permits the formulation of compatible common frames of reference and a customized glossary of terms to serve the needs of the project.

ORGANIZING AROUND THE RESEARCH PROBLEM

A serious error made by virtually every funding source and by recipient interdisciplinary research teams is not allowing enough start-up time to ponder, appraise and delineate the nature of the research problem. Considerable attention has been directed to the problem by researchers and administrators in the U.S.D.A. as well as Agricultural Experiment Stations. For new regional research projects dealing primarily with rural development problems the first year is designated and treated as "temporary committee" time. This means there is administrative recognition that start-up time is as important for research as the warm-up period for athletes or the count-down procedure for launching space vehicles.

Team members found that defining the scope of the project and the specific research objectives were much more difficult than originally envisaged. A first emphasis was centered on nitrogen as a runaway nutrient in the food production process. The attempt to focus on nitrogen as a problem assumed that it was a national concern and that it was also of principal concern in the Northeast. After working approximately one year on the project the interdisciplinary value was dramatically highlighted when the aquatic scientists brought forth data showing that for lakes in the temperate climate of the northeastern United States nitrogen is not the critical nutrient which controls the degree of eutrophication. At the watershed level the nutrient phosphorus was alternatively identified as the nutrient of concern. Measuring and managing phosphorus posed a rather different set of problems for the research team.

It was clear that this problem had not been as intensively studied and that for the watershed new empirical data would be required. How to organize this research undertaking was a subject of considerable debate as the various members of the team each had different opinions of what they or others might do.

Much of the early behavior of the interdisciplinary team can be attributed to the uncertainty the members felt about pinning down the nature of the research problem. The definition, delineation and explication of the research problem has many troublesome features. Each disciplinary representative has a tendency to view the problem from the perspective of his background and experience, thus being better able to describe and articulate its nature from the vantage point of his understanding and conceptual framework. It is, in fact, within this context that many of the earlier-mentioned problems of language, terminology, definition, meanings, standards and units of analysis became the subject matter of verbal jousting. The ability to control the temperament of a group as it moves through this process is a particular challenge, not only for a group leader but for those members who maintain a sense of humor and carry out some of the peace-keeping and maintenance roles required.

At a conceptual level, virtually every participant would agree that the basic elements of the research process from beginning to end consist of the following steps: (1) a definition of the problem, (2) conceptualization and identification of a series of hypotheses about the nature of the problem, (3) development of plans, procedures and necessary instruments to collect data and/or information relative to the hypotheses, (4) collecting the necessary data, (5) developing methods and procedures to analyze and interpret the data, and (6) publication, dissemination and application of research findings about the problem.

While agreement is achievable on the basic points outlined above, researchers differ in their experience, logic and perceived feasibility of starting at differing stages of these processes. Some wish to use a pure inductive approach while others prefer either a deductive procedure or a mixture of the two. Some find it easier to begin work by addressing one stage, some another. The tendency of researchers from various disciplines is to move by starts and stops, to move unevenly and for some to move far out in front while those highly dependent upon data sources being developed by others lag behind.

To a considerable extent, each researcher feels pressure from within his discipline to satisfactorily work through the research stages both theoretically and methodologically before it is reasonable to expect substantial integrative inputs into the content and findings of other disciplines. All of this suggests that a broader more flexible frame of reference needs to be employed in managing an interdisciplinary group's function than would be the case of researchers working within a single discipline (Capener, 1973).

The element of professional peer pressure was well illustrated by the experience of the interdisciplinary team in delineating a whole watershed as the unit of analysis for the study. The problems posed were especially difficult for the agronomists. They had been accustomed to working with small plots where the parameters could be brought more easily under controlled conditions. The prospect of sampling and measuring runoff from all of the land in a watershed posed a challenge of an entirely different magnitude. This meant moving from a limited number of measurement stations to a series of sampling sites that would indeed reflect nutrient runoff patterns in a whole watershed. It meant developing a design to acquire regularized measurements of a periodic nature, but also required data on the significance of critical event measures such as heavy rainstorms or snowfalls. It called for staff to move on short notice to gather samples and to bring them into the laboratory at odd hours since rapid analysis was crucial to preventing changes in the phosphorus chemistry. It was important to be able to compare and equate results from this critical event sampling procedure against that employed in the more routine sampling pattern.

As sample data were examined, they generated anxiety and pressure to make the right interpretations. The difficulty of explaining the uneven results was further compounded by the lack of comparative data since most other studies were based on routine time samples rather than on those taken during critical events. The challenge of setting forth interpretative results posed some risk to one's professional reputation, especially when it called for dealing with data obtained on a macroscale of a watershed as against the previously accustomed, more easily tested and corroborated microscaled data.

Given the tremendous variation in the sources of phosphorus input in a stream from a total watershed, the ability to assign proportional loading from the various sources posed a problem of severe magnitude. What was the importance of forest and uncultivated areas, of the heavily cropped sectors of the watershed and of the more dense residential portions? What was the proportional input from the primary point source, the sewage treatment plant? Each of these sources yielded data of considerable variability in terms of time-based samples, as compared to crisis event samples.

The agricultural economists and aquatic scientists were similarly faced with complex decisions about available and relevant data, relating various forms and amounts of phosphorus in water to the degrees of eutrophication, fitting the data to the scale of the watershed and developing interpretations that were defensible from a professional and scientific point of view. The rural sociologists gathered a new set of empirical data from a significant sample of residents in the watershed. Their problem was one of adequately interpreting individual, institutional and local governmental interest and capability of response to management options. The engineers concentrated on the problem of nitrogen removal from animal wastes in relation to its impact on the watershed.

As each disciplinary subgroup wrestled with the basic elements of the research process and the stages and phases emerged into a clearer perspective, it became easier for one subgroup to share its work and progress with the others.

THE TIME FACTOR IN
INTERDISCIPLINARY RESEARCH

An important need of an interdisciplinary group is to acquire an early sense of direction and accomplishment for the work and efforts contributed. This was facilitated to some extent by the use of agendas and minutes to create meeting expectations, a group memory and a record of progress. Other options were to establish divisions of labor, special assignments, and highlighting new areas and next phases to be addressed.

Normally a research group will be highly task-oriented; therefore the members will be receptive to assignments and divisions of labor. This was true of the interdisciplinary group. They accepted assignments and met regularly. The most marked sense of cohesion came to the interdisciplinary team in two stages. The first was when the decision was reached to study the whole watershed. The second was when each discipline subgroup had completed sufficient research to formulate early drafts for others' inspection and analysis. The drafts served to highlight common themes that would characterize the report. At this stage the points of data disagreement, questions of interpretation and blending of results could more easily be resolved. The prominent fabric of the study design could begin to take shape. The team agonized long and hard over the significance of the various chemical forms of phosphorus in water. The final agreement between social scientists, aquatic scientists, economists, engineers and sociologists was a significant accomplishment of the project. The group created a logo that could portray in picture form the interrelatedness of the disciplines in the study. The logo came to symbolize itself a growing unity in the research team.

As the final shape of the research report began to emerge it became more evident to the participants that a significant and valuable report could be written. The study contained original data that had not been previously presented, especially in a manner related to policy implications as derived from the interdisciplinary overview.

Regarding the time element in research involving several disciplines, there is no question that it takes more time the first time. The reason is for the needful processes of group formation and development to occur on the one hand and to work through the basic elements or steps in the research process on the other. These two major requirements superimposed one upon the other serve to create predictable obstacles. If the processes are understood, however, as they largely were by the leader and several members of the team, then the time line lends itself to management skills and a group can become relatively sophisticated in its abilities to interface disciplines around research projects with enhanced results. This fact is widely borne out by private contract research organizations who undertake projects where the leadership, division of labor, units of analysis, types of data, interpretations, publications and style of dissemination are carefully considered, decided and implemented.

The issues of academic freedom versus productivity are readily resolved. If there is no product there is no job. In another context a number of government agencies have the semblance of interdisciplinary staffing. This may explain certain program constraints, but on the other hand the personnel do in fact learn how to produce results or face dwindling support.

The evidence was conclusive for the interdisciplinary research team that once a team becomes integrated, the shorter will be the time required to tackle additional projects. In addition, once the results begin to appear, the importance of the findings and their enhanced relationships yield greater relevance and meaning to real world problems.

SPECIAL OBSERVATIONS

As might be expected there were several instances of directions being pursued to no avail, plans being abandoned for lack of feasibility or utility. For example, there was an early proposal to develop a scheme whereby the interactions of phosphorus could be traced and modeled. The naive expectation here was that each discipline could produce a set of calibrated numbers that could be incorporated into a comprehensive model. Data do not exist for variables in a model such as social costs, incentives for compliance and predispositions of farmers or residents to respond to controls.

Some of the more promising early attempts to work across discipline lines, to pursue common definitions, to seek common conceptual frames of reference and to move toward the interdisciplinary approach were made by the research associates under the direction of an assistant professor who was requested to investigate a mathematical model for the watershed. Perhaps since the status positions of this group were less pressured, they were able to direct attention to the larger interactive processes in the watershed. Later the research associates drafted much of the material that provided an integrative framework for the final report. While the model was never completed, partly due to lack of data, the meetings of this group were of great value to the participants in the project.

Other interesting points of concern arose from the seeming success of the project. These were questions about equitable distribution of the elements of recognition and reward. How should one go about listing the contributions various individuals had made to the project? If the junior member did more writing, should he or the senior partners be listed first? If persons had done more of the research and less of the writing, should they appear as joint authors or as a part of an acknowledgment section? If young staff are concerned about recognition and publication credits to enhance their career developments, how much is it worth if they are listed as one among a number of joint authors?

Another question which the fruits of interdisciplinary research raise is the identification of the different types of audiences or clients for whom the project findings are relevant. In a single discipline report, the audience identification is generally clear-cut. An interdisciplinary report evokes a wider range of potential clients, and therefore more complex publication issues need to be resolved.

The task of editing an interdisciplinary report is also more difficult. There is the style of writing to be settled and this is largely influenced by the nature of audiences to whom the publication is directed. Since several disciplines have to be represented and interpreted, an editor needs to be commissioned. Then there is the matter of satisfying the individual scientist within the disciplines that an editor has not done violence to his technical material such that it loses intended meaning or impact. These processes require an extra amount of time that is seldom adequately planned for at the outset. It could be said, then, that interdisciplinary research not only takes longer to carry out, but it also takes a longer period of time to finalize, publish and disseminate than is normally anticipated in terms of an individual scientist's time budget. While the interdisciplinary procedure appears slow, it probably represents the only means by which the broad objectives of this project could have been achieved.

A challenging complexity is posed in interdisciplinary research when it comes to addressing the potential policy implications of the findings. A single discipline may be able to draw out such policy statements with relative ease. The nature of policy statements that will hold across the disciplines, however, will generally be stated at higher levels of application and political or administrative jurisdiction.

In summarizing the advantages and disadvantages of interdisciplinary work along with highlighting some of the processes, the following considerations are noted:

A. Advantages

1. The foregoing elements of the interdisciplinary research process, taken singly or collectively, provide the participants with an opportunity for a comprehensive and, for many, a unique view of the kind of research needed to solve complex, real world prob-

lems. Further, the relatively small size of the group insured that all members are challenged to take part in any or all phases of the project from budgetary planning to the most complicated discussions of technical material. Participation of this kind adds a new and potentially valuable element to the training for research participants which several have gratefully acknowledged, *i.e.,*

> The experience of working on an interdisciplinary research project has increased my appreciation for the complexity of such an approach. It has provided an opportunity to become acquainted with the principles and concepts of other disciplines which has broadened my understanding of education. I believe the final product of this research effort will be of greater value because of the hard struggle the team has gone through. At this point I can say I am glad that I was a part of it.
>
> > William Schaffner
> > Research Associate

> The interdisciplinary experience has provided a unique and valuable training experience for me in research organization and administration. I foresee that much of my future professional career in research will necessarily involve other disciplines. I've gained a great deal of insight and understanding about what to emphasize, what to avoid and in general how to proceed.
>
> > William Saint
> > Graduate Assistant

> Participation in the project broadened my understanding of the nutrient management problem. Further it provided data from several disciplines which made it possible to develop my thesis. I feel the experience was very rewarding.
>
> > Thomas Tseng
> > Graduate Assistant

2. It is hard to visualize a better situation for leadership training and experience than an interdisciplinary research group. This is one of the very tangible benefits obtained from such projects. The task of the leader is to create a team from a disparate group of individuals representing different disciplines and made up of graduate students, postdoctoral researchers and faculty each having unique commitments to and interests in the project as a whole. Effective leadership training is also necessary at the subgroup level if each disciplinary segment of the research is to achieve the desired results within the time frame and be compatible with the project as a whole.

3. The overall results of an interdisciplinary project will be pre-

sented in a final report which should reflect more breadth and depth and be more useful and more applicable than a single disciplinary approach which can only address a smaller portion of the problem.

4. The results of an interdisciplinary project are potentially of interest to a wider audience.

5. An interdisciplinary project will provide spin-off utility in a wider range of publications.

6. The summary and implications of an interdisciplinary project will more likely generate a broader set of policy recommendations.

7. The members of a successful interdisciplinary project will have broadened their own insight, understanding and ability to work across disciplinary lines. For those professors who are teaching, much of value will be carried over into the classroom.

8. Once the interdisciplinary team has successfully worked on a first project, it will be infinitely easier for them to move ahead faster and more productively on a second experience.

9. A department, college or institution gains significant stature in its recognition and ability to handle future complex problems, once it has successfully demonstrated a capability of carrying out interdisciplinary research.

B. Disadvantages

1. Interdisciplinary projects generally take longer to carry out as compared to those of a more focused single discipline nature.

2. The problems of definition, delineation and explication of the research problem are more complex and difficult in interdisciplinary research.

3. Severe initial problems are posed for members of an interdisciplinary team to understand concepts, terminology, meanings, standards and differences in research procedures utilized between and among each other.

4. Strong challenges are placed against one's professional reputation, status image and feelings of support and competency.

5. It is difficult to find a person with experience and skill to provide leadership to an interdisciplinary group.

6. There are complicated problems of providing adequate recognition and rewards to encourage members to move through the various processes and stages to achieve success in an interdisciplinary team.

7. It is difficult to provide feedback or awareness to members of an interdisciplinary group on the nature of the growth and development processes they are experiencing.

8. The problems of equitable recognition and distribution of rewards in terms of professional publications are more complex in interdisciplinary research and are of major concern to the younger participants in terms of career credits.

9. There is always the threat of an interdisciplinary group blowing apart and not achieving a successful experience. A bad experience with interdisciplinary research is likely to last a long time.

REFERENCES

Capener, H. R. 1973. Social science research on water resource issues: problems of organization, administration and management. In W. H. Andrews, R. J. Burdge, H. R. Capener, W. K. Warner and K. P. Wilkinson, eds. The Social Well-Being and Quality of Life Dimension in Water Resources Planning and Development. Institute for Social Science Research on Natural Resources, Utah State University.

Fishel, W. L., ed. 1971. Resource Allocation in Agricultural Research. University of Minnesota Press. Minneapolis.

Hare, P. A., E. F. Borgatta, and R. F. Bales, eds. 1955. Small Groups: Studies in Social Interaction. Alfred A. Knopf. New York.

Harris, M. and R. J. Hildreth. 1968. Reflections on the organization of regional research activities. Amer. Jrn. Agr. Econ. 50:4, pp. 815–826.

Knoblaugh, H. C. et al. 1962. State Agricultural Experiment Stations: A History of Research Policy and Procedure. U.S. Department of Agriculture Mis. Pub. 904. Washington, D.C.

Lee, A. T. M. 1969. Regional research further analysis and reflections. Amer. Jrn. Agr. Econ. 51:4. pp. 953–957.

U.S. Department of Agriculture. 1963. Manual of Procedures for Cooperative Regional Research. CSESS (CSRS)-OD-1082. Washington, D.C.

ACKNOWLEDGMENTS

Grateful acknowledgment is made to each of the interdisciplinary research team members for their valuable insights and suggestions. The success of this project was only made possible through their fund of goodwill, patient cooperation and sense of loyalty to each other and to the project.

Principal Authors

Harold R. Capener
Robert J. Young

CONVERSION TABLE

To convert column A to column B, multiply column A by	Column A	Column B	To convert column B to column A, multiply column B by
Length			
0.621	kilometer (km)	mile (mi)	1.609
1.094	meter (m)	yard (yd)	0.914
3.282	meter (m)	foot (ft)	0.305
0.394	centimeter (cm)	inch (in)	2.54
Area			
0.386	kilometer^2 (km^2)	mile^2 (mi^2)	2.590
2.471	hectare (ha)	acre	0.405
0.0001	hectare (ha)	meter^2 (m^2)	10,000
Volume			
0.0042	meter^3 (m^3)	gallon (gal)	236.8
0.001	meter^3 (m^3)	liter (l)	1,000
1.057	liter (l)	quart (qt) (liquid)	0.947
4.223	liter (l)	gallon (gal)	0.2368
1.0940	meter^3 (m^3)	feet^3 (ft^3)	0.9141
Mass			
1.102	metric ton (mt)	ton (t) (English)	0.9072
10^6	gram (g)	microgram (microg. or μg)	10^{-6}
2.205	kilogram (kg)	pound (lb)	0.454
1,000	gram (g)	milligram (mg)	.001
Yield, Rate or Concentration			
0.446	metric ton/hectare (mt/ha)	ton (English)/acre (t/acre)	2.240
0.892	kilograms/hectare (kg/ha)	pounds/acre (lb/acre)	1.12
22.8	meter^3/second (m^3/sec)	million gallons/day	0.0438
35.3	meter^3/second (m^3/sec)	feet^3/second (ft^3/sec)	0.0283

GLOSSARY

Aerobic bacteria: Bacteria which require the presence of free (dissolved or molecular) oxygen for their metabolic processes.

Algal bloom: A concentrated growth or aggregation of phytoplankton that is readily visible.

Alluvial: Specifying material deposited from streams.

Ammonification: The biological process in which ammonium is formed from organic compounds.

Anaerobic bacteria: Bacteria which do not require the presence of dissolved or molecular oxygen for their metabolic processes.

Anoxic: With respect to an organism, conditions in which levels of oxygen are inadequate to sustain that organism.

Autotrophs: Bacteria which utilize carbon dioxide for cellular carbon.

Biogeochemical phosphorus: That phosphorus in stream discharge which is derived from sources not managed or appreciably influenced by human activity.

Biologically available phosphorus (BAP): An operational term used for the sum of all those forms of phosphorus entering a lake which rationally would appear to be available for biological uptake. These are molybdate reactive phosphorus, soluble unreactive phosphorus, and labile phosphorus.

Biomass: The weight of all life in a specified unit of an environment, which may be defined in terms of components such as a population, community, or as the total biota.

Biota: All animal and plant life in an aquatic or terrestrial system.

BOD (biological oxygen demand): An indirect measure of the concentration of biologically degradable material present in organic wastes. It is the amount of free oxygen utilized by aerobic organisms when allowed to attack the organic matter in an aerobically maintained environment at a specified temperature (20°C) for a specific time period (5 days), and is expressed in mg oxygen utilized per liter liquid waste volume.

Chlorophyll a: The primary photosynthetic pigment of all plants.

COD (chemical oxygen demand): An indirect measure of the oxygen demand exerted on a body of water when organic wastes are introduced into the water. It is determined by the amount of potassium dichromate consumed in a boiling mixture of chromic and sulfuric acids. The amount of oxidized organic matter is proportional to the potassium dichromate consumed. Where wastes contain only readily available organic bacterial food and no toxic matter, COD can be the equivalent of the BOD obtained from the same wastes.

Denitrification: The biological reduction of nitrate and nitrite with the liberation of molecular nitrogen and in some instances, gaseous nitrogen oxides.

Discharge rate: The amount of water discharged during specified time periods.

Drainage basin: The total area that drains into a lake or other body of water.

Enrichment ratio: The concentration of a nutrient in eroded soil relative to the concentration of the nutrient in the topsoil from which the eroded soil was derived.

Epilimnion: The layer of water above the thermocline (or above the metalimnion) of a lake. Characterized by complete mixing, and is that portion of the lake in which most or all of the photosynthesis occurs.

Euphotic zone: In lakes, the depth zone which is penetrated by sufficient light to permit net photosynthesis (the growth of green plants).

Eutrophication: The enrichment of a body of water with plant nutrients.

Evapotranspiration: The process of transferring moisture from the earth to the atmosphere by evaporation of water and transpiration from plants.

Facultative bacteria: Bacteria which can exist and reproduce under either aerobic or anaerobic conditions.

Feedlot: A confined operation for the feeding of livestock, the conditions of which preclude the growth of vegetation.

Glacial outwash: Glacial material which has been moved and sometimes sorted by water movement.

Glacial till: The unsorted material (clay, silt, sand, boulders) deposited by glaciers.

Gutter flush: A system by which manure is flushed from the gutter of a stanchion-type barn by the addition of a small amount of water to make the manure flow.

Heterotrophs: Bacteria which use organic carbon for their carbon source.

Hydraulic retention time: The average length of time that a given particle of water will spend in a tank.

Hypolimnion: The layer of water below the thermocline (or below the metalimnion), characterized by low temperatures, little or no light, and possible oxygen deficiencies in certain lakes in the latter part of the summer.

Infiltration rate: The rate at which water can enter the environment of the soil through its surface, usually expressed in inches of water per day.

Kjeldahl nitrogen: Includes organic nitrogen and ammonia but not nitrite- and nitrate-nitrogen, and is determined using a method developed by J. Kjeldahl.

Lake retention time: The average length of time that a given particle of water will spend in a lake.

Littoral: Pertaining to the shallow area of a lake.

Load (stream): The quantity of some specified substance (e.g., sediment, nitrate-nitrogen, or phosphorus) carried in stream discharge over a specified time period.

Loading (waste stabilization system): Two types of loading may be distinguished: organic and hydraulic. The former refers to the rate at which organic mass is applied to the system, and the latter to the rate of flow of the wastewater being treated.

Metalimnion: The layer of water encompassing the thermocline and characterized throughout by a decrease of temperature with depth. Bounded above by the epilimnion and below by the hypolimnion.

Methemoglobinemia: A physiological condition characterized by the presence of methemoglobin in the blood which reduces its oxygen-carrying capacity.

Mineralization: The process by which organic compounds are transformed, usually by bacteria, to inorganic substances.

Mixed liquor: Usually refers to the combination of wastewater and microbial flocs or sludges produced during waste treatment.

Mixed liquor volatile suspended solids (MLVSS): A measure of the active organisms in an aeration tank.

Molybdate reactive phosphorus (MRP): See Table 2.1.

Nitratification: A process of oxidation of nitrite to nitrate (NO_2^- to NO_3^-).

Nitrification: The biological formation of nitrite or nitrate from compounds containing reduced nitrogen.

Nitritification: A process of oxidation of ammonium (NH_4^+) to nitrite (NO_2^-).

Nitrogen fixation: Any process of combining atmospheric nitrogen with other elements; the process performed by certain bacteria in the nodules of leguminous plants, which make the resulting nitrogenous compounds available to their host plants.

Non-point source: A nonspecific, unidentified, or diffuse source from which organic and inorganic materials enter surface and ground water. Effectively, any source not defined as a point source may be considered a non-point source.

Oligotrophic: Characterized by low plant productivity and water with high transparency; waters that are nutrient poor.

Oxidation ditch: An aerobic biological waste treatment system, with long liquid detention time and adequate mixing by a surface aerator.

pH: Used to indicate an acid or alkaline condition (pH 7 indicates neutral; less than 7 is acid; greater than 7 is basic).

Phytoplankton: Free-floating and usually microscopic algae.

Point source: Any discernible, confined and discrete conveyance, including but not limited to any pipe ditch, channel, tunnel, concentrated animal feeding operation, or vessel, from which pollutants are or may be discharged.

Primary sewage treatment: A method of treatment, the main purpose of which is to remove the larger, suspended particles of organic matter by gravitational settling.

Secchi disc: A reflective object which can be lowered into the water on a calibrated line to measure transparency. *See* Figure 2.2.

Septic tank: A settling tank in which the organic portion of settled sludge is allowed to decompose anaerobically. Only partial liquefication and gasification of the organic matter is accomplished, and eventually removal of undecomposed solids is necessary.

Secondary sewage treatment: Following primary treatment, a biological stage utilizing microorganisms to further the stabilization of organic materials in solution and fine suspension.

Sorption isotherms: The relationship between concentration of phosphorus in solution and amount of sorbed phosphorus on suspended solids in water. *See* Appendix A, Chapter 3.

Soluble unreactive phosphorus (SUP): *See* Table 2.1.

Specific phosphorus loading: *See* Table 2.1.

Stabilization (biological): Oxidation of organic material as a result of the metabolic activity of organisms.

Standing crop: The biomass of an organism or group of organisms present in a lake at any given time, expressed per unit volume or as that amount under a unit area of lake surface.

Substrate: A media used to maintain or grow organisms.

Supernatant: The clear liquid remaining after solids separation.

Tertiary sewage treatment: A method to further purify sewage that has received secondary treatment.

Thermocline: The plane in a column of water at which the temperature decrease is greatest.

Total phosphorus (lake, TP_L): *See* Table 2.1.

Total phosphorus (stream, TP_s): The sum of the total soluble phosphorus and the phosphorus in the suspended solids of a stream sample.

Total soluble phosphorus (TSP): The total amount of phosphorus in the supernatant of a stream sample after oxidation with persulfate. *See* Appendix A, Chapter 3.

Trophic state: Literally, the nutritive state of a lake. Often judged by the standing crop of plants (*e.g.,* a lake in a eutrophic state would be rich in nutrients and have a dense population of phytoplankton or rooted aquatic plants).

Unconsolidated mantle: The clay, silt, sand, gravel, boulders, etc., overlying the solid bedrock.

Volatile solids: That portion of the total or suspended solids residue which is driven off as volatile (combustible) gases at a specified temperature and time (usually at 600°C for at least one hour).

Volatilization: The process of vaporizing or becoming gaseous.

Watershed: The area drained by a stream and its tributaries.

INDEX